W9-BDV-734

Global Warming, Natural Hazards, and Emergency Management

Global Warming, Natural Hazards, and Emergency Management

Jane A. Bullock

George D. Haddow

Kim S. Haddow

CRC Press
Taylor & Francis Group
Boca Raton   London   New York

CRC Press is an imprint of the
Taylor & Francis Group, an **Informa** business

Cover photos: Top row: left image courtesy NOAA Photo Library, center image courtesy NOAA Photo Library, NOAA Central Library; OAR/ERL/National Severe Storms Laboratory (NSSL), right image courtesy Greg Mathieson/ FEMA News Photo. Middle row: center image courtesy Greg Henshall/FEMA, right image courtesy NOAA Photo Library. Bottom row: left image courtesy Jocelyn Augustino/FEMA, center image courtesy Andrea Booher/FEMA News Photo, right image courtesy Marvin Nauman/FEMA.

CRC Press
Taylor & Francis Group
6000 Broken Sound Parkway NW, Suite 300
Boca Raton, FL 33487-2742

© 2009 by Taylor & Francis Group, LLC
CRC Press is an imprint of Taylor & Francis Group, an Informa business

No claim to original U.S. Government works
Printed in the United States of America on acid-free paper
10 9 8 7 6 5 4 3 2 1

International Standard Book Number-13: 978-1-4200-8182-4 (Softcover)

This book contains information obtained from authentic and highly regarded sources. Reasonable efforts have been made to publish reliable data and information, but the author and publisher cannot assume responsibility for the validity of all materials or the consequences of their use. The authors and publishers have attempted to trace the copyright holders of all material reproduced in this publication and apologize to copyright holders if permission to publish in this form has not been obtained. If any copyright material has not been acknowledged please write and let us know so we may rectify in any future reprint.

Except as permitted under U.S. Copyright Law, no part of this book may be reprinted, reproduced, transmitted, or utilized in any form by any electronic, mechanical, or other means, now known or hereafter invented, including photocopying, microfilming, and recording, or in any information storage or retrieval system, without written permission from the publishers.

For permission to photocopy or use material electronically from this work, please access www.copyright.com (http://www.copyright.com/) or contact the Copyright Clearance Center, Inc. (CCC), 222 Rosewood Drive, Danvers, MA 01923, 978-750-8400. CCC is a not-for-profit organization that provides licenses and registration for a variety of users. For organizations that have been granted a photocopy license by the CCC, a separate system of payment has been arranged.

Trademark Notice: Product or corporate names may be trademarks or registered trademarks, and are used only for identification and explanation without intent to infringe.

Library of Congress Cataloging-in-Publication Data

Global warming, natural hazards, and emergency management / editors, George Haddow,
Jane A. Bullock, Kim Haddow.
   p. cm.
Includes bibliographical references and index.
ISBN 978-1-4200-8182-4 (hardcover : alk. paper)
1. Global warming. 2. Natural disasters. 3. Emergency management. I. Haddow, George
D. II. Bullock, Jane A. III. Haddow, Kim.

QC981.8.G56G581943 2008
363.34'5--dc22
                                2008038326

Visit the Taylor & Francis Web site at
http://www.taylorandfrancis.com

and the CRC Press Web site at
http://www.crcpress.com

CONTENTS

FOREWORD

Ten years ago, for the first time in history, the directors of the Federal Emergency Management Agency (FEMA) and a major American environmental organization, the Sierra Club, sat down together. It took us no time to find common ground.

Former FEMA directors considered the agency's primary task to be to respond to natural disasters after they occurred. But James Lee Witt, FEMA Director in the Clinton administration, changed the focus of the agency from response to mitigation, giving priority to actions to reduce the risks and impacts of disasters before they happened. It was this shift in perspective and purpose that made the meeting between FEMA and the Sierra Club, between Witt and Carl Pope, possible.

Our meeting came on the heels of a series of disasters that had been exacerbated by compromised or devastated ecosystems. The Red River Flood had been worsened by the loss of wetlands. Lethal mudslides in Central America caused by deforestation were triggered by Hurricane Mitch and resulted in thousands of deaths. An unprecedented series of wildfires threatened communities in Florida.

In the course of our discussion, we quickly identified three common and closely held beliefs that connected our work and worlds:

- Protecting nature protects people. Wetlands, forest, barrier islands — all ecosystems form the first line of defense against natural disasters. They serve as a buffer against storm winds and as a sponge to soak up storm waters. Without them, communities are more vulnerable to disaster.
- Reducing risks before disaster strikes saves lives and money. Response and recovery are more expensive. Mitigation, reducing risks before a disaster, is cost-effective, saves lives, and prevents economic disruption. Not building in a floodplain, for example, will save lives and prevent property loss. Once development occurs, people and property have knowingly been placed in harm's way.
- Local communities have a large and unique role to play in reducing the risks of natural disasters. Local governments and local

leaders are on the front lines and are best positioned to establish disaster and mitigation planning processes. They also have the power to implement their plans by deciding zoning and land-use issues, setting building codes and standards, and overseeing the location, development, and maintenance of roads, bridges, and other infrastructure.

We parted in 1998 with a handshake and promise to work together inside and outside the administration to persuade and pressure the Army Corps of Engineers to change course and to make it harder for developers to pave over wetlands. A decade later, we are coming together again, this time driven by the urgent need to prepare communities for the inescapable effects of climate change.

The science is clear — the climate is changing. The planet is heating up, we are already experiencing the effects, and it will get worse before it gets better. We are living with the consequences of climate change — temperatures are on the rise, glaciers are melting, snowpack is disappearing, sea levels are rising — all changes that increase the risk of floods, droughts, and wildfires.

The internationally recognized authority on global warming, the Intergovernmental Panel on Climate Change (IPCC) has warned us that we should brace for more extreme weather events and that more natural disasters are inevitable. And even if we succeed in dramatically reducing emissions of carbon dioxide and the other greenhouse emissions that cause global warming, decades of global warming are inevitable — past emissions will continue to warm the earth and the effects of that are inescapable

Bottom line: communities accustomed to seasonal floods, droughts, and wildfires are likely to experience more frequent and intense floods, droughts, and wildfires. And we can expect the range of disasters to expand — areas that were never touched by extreme weather will be affected because of climate change.

So, while it is essential to cut carbon emissions, to switch from our dependence on the fossil fuels that emit greenhouse gases to clean, renewable solar and wind power, and to increase the energy efficiency of our buildings, vehicles, and appliances, it is not enough to address only the cause side of the equation. We need to act urgently to cut emissions and prepare for the consequences of global warming, and local leadership is essential to both.

As our friend Ron Sims, Washington's King County Executive, says, we cannot afford the luxury of not preparing, because some impacts are inevitable: "We must prepare for the impacts under way while we work to avoid even worse future effects."

In her remarks to the American Association for the Advancement of Science, Judith Rodin, President of the Rockefeller Fund, echoed the need to increase focus on adaptation: "Since it may be too late to stop global warming that's already occurred, we must focus on how to survive it. Currently there is less attention paid in the scientific and policy communities to adaptation, [to] what needs to be done to help people and environments cope with what's already occurred and what's coming."

So, how do communities adapt and mitigate more extreme heat waves, storms, floods, water shortages, coastal erosion, and all the other consequences of global warming? They come together as a community, identify their risks, and develop strategies to reduce those risks. And they build on what has already succeeded.

Although the federal government has a role to play in providing financing, incentives, and support, it is local communities that are on the front lines where climate change impacts are felt most directly, and it is local communities that are best able to assess and tailor their mitigation and adaptation efforts to the local and regional threats created by the changing climate. It is the job of local communities to institute water conservation programs, to restrict or prevent building in the floodplains, to restore wetlands, and to educate their citizens.

This book provides local governments with replicable case histories — and hope. Included here are success stories, stories about the communities of Napa, California, and Grand Forks, North Dakota, which have reduced their flood risks, of Tulsa, Oklahoma, which worked to reduce the toll of tornadoes, and western towns that are taking steps to fight wildfires by creating Community Buffer Zones in their wilderness interface. But most importantly, it offers a process and resources for disaster planning at the community level that have been proven successful and have never been needed more urgently.

In the decade since we last met, the scale, immediacy, and intensity of the challenges we face have changed radically. But the three fundamental principles we recognized as primary drivers in our overlapping worlds still stand and inform this book: the need to protect and restore the natural systems that are the first line of defense against natural disasters;

the importance of reducing the risks of future disasters, not just being able to respond to them; and that local communities must take the lead.

Finally, even though the scientists tell us it is too late to avoid climate change, it is not too late to make it a smaller problem for our children and grandchildren. Our hope is that this book will make it easier for communities to act now.

Carl Pope
Executive Director
Sierra Club

James L. Witt
Chief Executive Officer
James Lee Witt Associates, Inc.

ACKNOWLEDGMENTS

First and foremost, we want to acknowledge the contributors to this book: Arrietta Chakos, Damon Coppola, Brian Cowan, Dave Dickson, Richard Gross, Kathryn Hohmann, Fran McCarthy, Ann Patton, Ines Pearce and Jim Schwartz. It is their stories that are the heart of this book and their dedication to protecting their communities and fellow citizens that is a lesson to us all.

Secondly, many of the recommendations presented in our concluding chapter were first presented in a paper entitled, "Forecast: Storm Warnings – Preparing for More Severe Hurricanes Due to Global Warming" that we co-authored with Kit Batten, Benjamin Goldstein, Bracken Hendricks, Kari Manlove, and Daniel J. Weiss for the Center for American Progress in Washington, DC.

Finally, we want to dedicate this book to James Lee Witt who has set the standard for leadership in public service.

INTRODUCTION

Efforts to slow and reverse climate change will take at least three to four generations. In the interim, scientists predict that the frequency and severity of weather-related disaster events will increase dramatically. So are there actions that can be taken now that will reduce the impacts of these future disasters intensified by climate change on individuals, communities, economies, and the environment? The answer is yes. Programs designed to reduce the risks and impacts of natural hazards, when implemented, have succeeded in saving lives and property and have demonstrated collateral benefits in reducing the impact of climate change on our communities.

This book identifies what has worked to mitigate natural hazards in communities across America and examines how to apply those lessons to help us increase our defenses and reduce the impact of the effects of a changing climate.

Mitigation is a word that straddles two worlds, and it is a concept that can help translate past efforts that have been successful in saving lives and property from natural disaster into a new context — a world where the changing climate is altering the intensity, frequency, and predictability of future disasters.

Steps to reduce risk and impacts in the world of natural disasters have been termed *mitigation. Mitigation* in the context of the global warming arena refers specifically to efforts to reduce the greenhouse gas emissions that are causing climate change. The meaning behind both of these notions is the same — that is, to take action now that reduces future consequences.

The goal of mitigation is the same in both worlds, whether it is to reduce the amount of carbon in the atmosphere so global warming is a smaller problem for future generations, or it is action taken now — restoring wetlands or banning development in a floodplain — to reduce the impact of future floods.

Just as both worlds agree on the need to mitigate, to act now to reduce future impact, both emergency managers and climate scientists advocate preparing for the inevitable, for the impacts that will come with the next storm, drought, or wildfire.

Until now, the idea of preparing for the inevitable change caused by global warming has been controversial. Some scientists, policy makers, and activists fear that "adapting" to the changing climate diverts attention and resources needed to reduce greenhouse gas emissions. But now, a new wave of thinking is forcing reconsideration and recognition that both cause and consequences must be addressed. It is a lesson the disaster management world absorbed and applied two generations ago with the creation of the National Flood Insurance Program in 1968. It is a lesson we must all embrace today.

According to the *Christian Science Monitor*, "Ironically, many measures needed to adapt to global warming come from the same toolkit disaster planners and development agencies use today. 'Adaptation means doing the things you do now, but doing them better,' says World Bank Climate Change Specialist, Dr. Ian Noble."[*]

The purpose of this book is to present a series of essays and case studies of current and past hazard mitigation efforts that have been successful in reducing disaster impacts. These essays and case studies have been authored by individuals who were directly involved in the successful design and implementation of community-based hazard mitigation programs. Collectively, these essays and case studies provide a clear procedural road map for emergency managers, policy makers, and community officials on how to reduce the impact of future disaster events that are being intensified by the effects of global warming.

Chapter One examines the most current thinking in the scientific community on climate change and how to best address the problem. For years, the belief among scientists and policy makers was that mitigating the causes of global warming (i.e., reducing emissions, etc.) was the single most important action to be taken to reduce, reverse, and eliminate global warming. Today the consensus in the scientific community is that the consequences of climate change are inevitable and that reducing the impact of global warming (e.g., more frequent and severe weather-related disasters such as drought, floods, hurricanes, etc.) through "adaptation" is of equal importance and that hazard-mitigation actions must occur in conjunction with efforts to mitigate the causes of global warming.

Chapter Two presents essays concerning the role urban and regional planners can play in community-hazard reduction and how the environment has been and will continue to be the first line of defense in protecting communities from a wide range of disasters influenced by global warming, including droughts, floods, and wildfires.

---

[*] Christian Science Monitor, "Time to Begin 'Adapting' to Climate Change?" February 13, 2007.

Chapter Three examines the wide range of hazard-mitigation programs sponsored by the Federal Emergency Management Agency (FEMA) that have played critical roles in addressing the increased frequency and severity of disasters caused by global warming. These programs include: the National Flood Insurance Program (NFIP), which provides flood insurance for communities and individuals under the condition that the community implements and enforces ordinances limiting certain development in the floodplain and precludes new development in special flood hazard areas; FEMA's Property Acquisition Program, which acquires flood-prone properties and removes them from harm's way; and Project Impact: Building a Disaster Resistant Community, which supported the establishment of community-based hazard mitigation programs in over 250 communities across the nation.

Chapter Four includes case studies of ongoing community-based hazard mitigation efforts in Tulsa, Oklahoma, and Berkeley, California. These case studies clearly illustrate the process that each community undertook to involve all community stakeholders in a partnership to address their hazard risks. The studies discuss the roles of government and private-sector officials and ordinary citizens in generating the political will to address these difficult issues and to secure funding for a series of projects designed to reduce the impact of future disasters.

Chapter Five offers examples of cross-jurisdictional risk-reduction efforts, including a description of how the communities in the Napa Valley in northern California conducted a two-year planning process that resulted in a 20-year plan designed to reduce the impact of flooding from the Napa River on the local residents, institutions, economy, and environment. A second case study details how officials from the government, private, and nonprofit sectors came together across state and international boundaries to create a series of programs designed to reduce flood impact in the Red River Basin in North Dakota, Minnesota, and Manitoba, Canada. A third case study examines how Seattle Project Impact designed and implemented programs across the region that protected home owners, schoolchildren, and businesses from the impact of earthquakes.

Chapter Six presents conclusions and recommendations based on the experiences and ideas presented in the essays and case studies. Those common features that can be found in each essay and case study are highlighted, along with the impact or role they had in the successful design and implementation of hazard-reduction programs. Based on these conclusions, a series of recommendations are presented concerning how officials and agencies in the federal, state, and local governments, the private sector and the nonprofit sector can support and promote programs in their communities that will reduce the impact of future disasters influenced by global warming.

The Appendix contains a listing of reports, Web sites, and other materials related to climate change and hazard-risk reduction.

Our hope is that the information presented in this book will make it clear to community leaders that there are successful models for building the types of community partnerships that will be needed to reduce the impact of future floods, droughts, wildfires, and other disasters influenced by global warming. It is to that end that this book is dedicated.

1

The Case for Adaptation (Risk Reduction)

Kim Haddow

Kim Haddow currently serves as the director of communications at the Sierra Club. She oversees the Club's branding efforts, strategic communications planning, message development, earned and paid media, and communication channels. Haddow joined the Sierra Club after working for nine years as the head of her own media consulting and advertising agency. Before starting her own business, Haddow spent eight years at Greer, Margolis, Mitchell, Burns — a consulting firm where she worked on 22 gubernatorial and senatorial candidate and statewide initiative campaigns. Haddow is a graduate of Washington College in Maryland and Loyola University of the South's Institute of Politics.

INTRODUCTION

The conclusion of the 2007 assessment by the Intergovernmental Panel on Climate Change (IPCC) could not have been starker: it is an "unequivocal" fact that the earth is getting hotter.[1] Climate change, according to the leading international network of climate experts, is real and its impacts are present, accelerating, intensifying, and inescapable.

The World Meteorological Organization (WMO) and the United Nations Environmental Programme (UNEP) created the IPCC in 1988 to assess "the latest scientific, technical and social-economic information relevant to the understanding of the risk of human-induced climate change, its observed and projected impacts and options for adaptation and mitigation."[2] The IPCC, which won the 2007 Nobel Peace Prize "for their efforts to build up and disseminate greater knowledge about man-made climate change, and to lay the foundations for the measures that are needed to counteract such change,"[3] represents the present scientific consensus and is considered to be the authoritative source on climate change.

The IPCC Fourth Assessment Report: Climate Change 2007 involved the work of 1,200 scientists and 2500 expert reviewers from 130 countries. It presented what the *New York Times* called a "bleak and powerful assessment of the future of the planet" and noted that a broad array of scientists consider it "the most sobering view yet of a century of transition — after thousands of years of relatively stable climate conditions — to a new norm of continual change."[4]

The Assessment listed both present, observed evidence that the earth's climate is changing and projected severe consequences of future climate change based on "greatly increased" number of studies and "improved" data sets[5] that allowed the group to be more specific and confident in its projections than in previous assessments. It also added new urgency to a debate that is now less focused on whether global warming is real and man-made, and more on how to reduce its causes and live with its consequences.

OBSERVED CLIMATE CHANGES

The IPCC assessed decades of climate data recorded from the depths of the oceans and miles above Earth's surface, and it concluded that climate change is "now evident from observations of increases in global average air and water temperatures, widespread melting of snow and ice and rising global average sea level."[6]

The report noted that eleven of the last 12 years rank among the 12 hottest years on record (since 1850, when sufficient worldwide temperature measurements began), concluded that most of the observed increase in globally averaged temperatures since the mid-20th century is very likely due to man-made greenhouse gas emissions and confirmed that the current atmospheric concentration of carbon dioxide and methane, two important

2

heat-trapping greenhouse gases, "exceeds by far the natural range over the last 650,000 years."[7]

The IPCC linked global warming to observed changes in climate, specifically, according to the analysis by the Union of Concerned Scientists, to:

Increasingly severe weather

- Increased precipitation in some areas: "From 1900–2005, precipitation increased significantly in eastern parts of North and South America, northern Europe, and northern and central Asia . . ."
- Increased drought in other areas: Droughts have become longer and more intense, and have affected larger areas since the 1970s, especially in the tropics and subtropics.
- Higher temperatures and more heat waves: "Average Northern Hemisphere temperatures during the second half of the 20th century were very likely higher than during any other 50-year period in the last 500 years . . ."
- More intense storms — The intensity of tropical cyclones (hurricanes) in the North Atlantic has increased since 1970.[8]

Melting and Thawing

- "Mountain glaciers and snow cover have declined worldwide.
- Since 1900, the Northern Hemisphere has lost seven percent of the maximum area covered by seasonally frozen ground.
- Satellite data since 1978 show that the extent of Arctic sea ice during the summer has shrunk by more than 20 percent."[9]

Rising Sea Levels

- Thermal expansion (ocean water expansion cause by absorbing the heat added to climate), melting glaciers, icecaps and the polar ice sheets have also contributed to recent sea level rise.[10]

Projected Climate Changes

The Temperature Continues to Rise

No matter what we do today, the earth's temperature will continue to climb. According to the IPCC report, "with current climate change mitigation and related sustainable development practices global greenhouse gas emissions will continue to grow over the next few decades."[11]

The Panel's projected climate change for the second half of this century was based on how much heat-trapping carbon, methane, and other greenhouse gases are emitted into the atmosphere. The IPCC based its projections on six different emission scenarios. The lowest temperature increase projected by the Panel for 2100 assumes a 2050 peak in world population, a rapid transition to service and information economy, and a shift toward clean and energy-efficient technologies. The highest temperatures projected for the end of this century assumes a mid-century peak in global population, rapid economic growth, and more "fossil intensive" energy production and consumption.[12]

Under any of the IPCC assessment scenarios, the Earth's temperature will continue to rise:

- The **full range** of projected temperature increase is 2 to 11.5 degrees Fahrenheit (1.1 to 6.4 degrees Celsius) by 2100.
- "The **best estimate range** of projected temperature increase, which extends from the midpoint of the lowest emission scenario to the midpoint of the highest, is 3.1 to 7.2 degrees Fahrenheit (1.8 to 4.0 degrees Celsius) by 2100."[13]

More Emissions, Higher Temperatures, More Climate Change

The evidence gathered and assessed by the IPCC indicates that a warming planet will cause intense and widespread devastation and disruption. "Continued Greenhouse Gas emissions at or above current rates would cause further warming and induce many changes in the global climate system during the 21st century that would very likely be larger than those observed during the 20th century."[14] Bottom line: We will see more of the same, only worse, more often, and in unexpected places.

The IPCC's project climate change and impacts, analyzed by the Union for Concerned Scientists, include:

Increasingly Severe Weather

- "Extreme heat, heat waves and heavy precipitation events will continue to become more frequent.
- Increases in the amount of high latitude precipitation are very likely, while decreases are likely in most subtropical land regions.
- Tropical hurricanes and typhoons are likely to become more intense, with higher peak wind speeds and heavier precipitation associated with warmer tropical seas."[15]

4

Melting and Thawing

- "Sea ice is projected to shrink in both the Arctic and Antarctic under all model simulations. Some projections show that by the latter part of the century, late-summer Arctic sea ice will disappear almost entirely."[16]

Sea Level Rise

The IPCC projects that sea levels will continue to rise, but "because understanding of some important effects driving sea level rise is too limited," the assessment does not offer "the likelihood, nor provide a best estimate or an upper bound for sea level rise."[17]

- "The models used by the IPCC project that by 2100, the global average sea-level will rise between 7 and 23 inches (0.18 and 0.59 meters) above the 1980–1999 average."[18]
- "Some models do suggest that sustained warming between 2 and 7 degrees Fahrenheit above today's global average temperature would initiate irreversible melting of the Greenland ice sheet — which could ultimately contribute about 23 *feet* to sea-level rise."[19]

Specifically, the consequences of these projections include:

- Heavy precipitation events, which are very likely to increase in frequency, will augment flood risk.
- Drought-affected areas will likely increase in extent.
- In the course of the century, water supplies stored in glaciers and cover are projected to decline, reducing water availability in regions supplied by melt water from major mountain ranges, where more than one-sixth of the world population currently lives.
- Approximately 20–30 percent of plant and animal species assessed so far are likely to be at increased risk of extinction
- Coasts are projected to be exposed to increasing risks, including coastal erosion, due to climate change and sea-level rise.
- Coastal wetlands including salt marshes and mangroves are projected to be negatively affected by sea-level rise, especially where they are constrained on their landward side, or starved of sediment
- Many millions more people are projected to be flooded every year due to sea-level rise by the 2080s. Those densely populated and low-lying areas where adaptive capacity is relatively low, and

which already face other challenges such as tropical storms or local coastal subsidence, are especially at risk

- Where extreme weather events become more intense and/or more frequent, the economic and social costs of those events will increase, and these increases will be substantial in the areas most directly affected.[20]

In North America, we can expect climate change will be different in different regions. Generally, we will experience:

- Decreased snowpack, more winter flooding, and reduced summer flows.
- Increasing competition for water resources.
- An extended period of high wildfire risk and large increases in area burned.
- An increased number, intensity, and duration of heat waves during the course of the century, with potential for adverse health impacts.
- Increasingly climate change–impacted coastal communities and habitats. Population growth and the rising value of infrastructure in coastal areas increase vulnerability to climate variability and future climate change, with losses projected to increase if the intensity of tropical storms increases.[21]

CHANGE GONNA' COME

Even if we stopped emitting all greenhouse gases today, warming and sea-level rise would continue to rise and the world would still be facing decades of climate change. The most optimistic scenarios of the IPCC project that concentrations of greenhouse gases will continue to climb: "For the next two decades a warming of about 0.2° C per decade is projected for a range of [emission scenarios]. Even if the concentrations of all greenhouse gases and aerosols had been kept constant at year 2000 levels, a further warming of about .1° C per decade would be expected."[22] That means past emissions will continue to heat the planet and their effects on the future are unavoidable and irreversible.

According to The Climate Impacts Group at the University of Washington and their partners in the King County Executive Office and at ICLEI-Local Governments for Sustainability, "Many of the climate changes projected through 2050 will be driven by present-day greenhouse gas emissions. Reducing greenhouse gas emissions will limit the severity of long term future impacts — but do little to change the near-term changes already set in motion."[23]

By midcentury, the climate in many areas of the United States will be significantly hotter than the warmest years of the last century and that temperature rise will increase the risks of floods, droughts, forest fires, and other disasters.

We are already seeing an increase in the number of natural disasters — from around 200 a year between 1987 and 1997, to double that between 2000 and 2006.[24] Floods are occurring more often, and they are affecting a larger land area than they did 20 years ago. Large-scale disasters — like the 2003 heat wave in Europe that killed 35,000 people and Hurricane Katrina, which caused $125 billion in damage in 2005 — are also happening with greater frequency.[25]

The amount of climate change that is projected for the next forty years will mean an increase in natural disasters — and not just in numbers but in severity and reach. According to Maarten Van Aaist, Associate Director and Lead Climate Specialist at the Red Cross/Crescent Climate Centre, "It aggravates the intensity and frequency of many hazards, but also creates surprises, such as hazards occurring in succession, or in places where they had never been experienced before. In terms of planning, past experience no longer guides what we can expect in the future."[26]

But according to the IPCC, warming and its effects can be substantially blunted by prompt action and planning. Achim Steiner, the executive director of the United Nations Environment Program, said society has the information it needs to act: "The implications of global warming over the coming decades for our industrial economy, water supplies, agriculture, biological diversity and even geopolitics are massive."[27] In the same New York Times article, Richard B. Alley, one of the lead authors and a professor at Penn State University, pressed policy makers to act with urgency: " . . . we have high very scientific confidence in this work — this is real, this is real, this is real. So now act, the ball's back in your court."[28]

MITIGATION AND ADAPTATION

Until now, the scientific and policy communities have made mitigation — reducing the amount of carbon and other greenhouse gases in the atmosphere — the primary focus of their efforts to stop global warming and stabilize the climate. It is a long-term attack on the cause of global warming that will take decades to achieve results and will do nothing in the short-term to help communities survive the climate changes that are coming. Concentrating solely on mitigation is no longer enough. "We must also focus on adaptation. Since it may be too late to stop global warming that's already occurred, we must focus on how to survive it . . . on what needs to be done to help people and environments cope with what's already occurred and what's coming," explained Judith Rodin, President of the Rockefeller Fund to the members of the American Association for the Advancement of Science.[29]

Historically, adaptation — increasing the ability of communities to survive and thrive in a warmer world — has been viewed an admission that we have given up and accepted climate change, that we do not need to address its causes by changing the way we produce and consume energy.

In an article published in the magazine *Nature*, "Lifting the Taboo on Adaptation," Daniel Sarewitz, director of Arizona State University's Consortium for Science, Policy and Outcomes, and his colleagues, argued, "Adaptation has been portrayed as a sort of selling out because it accepts that the future will be different from the present. Our point is the future will be different from the present no matter what, so not to adapt is to consign millions to death and disruption."[30]

In the same article, Sarewitz and his fellow policy experts reason, "The obsession with researching and reducing the human effects on climate change has obscured the more important problems of how to build more resilient and sustainable societies . . ."[31]

But thinking is changing as it becomes clear that, according to the IPCC, "There are some impacts for which adaptation is the only available and appropriate response." The need to include adaptation in a unified approach to surviving and solving climate change is becoming more widely adopted:[32]

> "As evidence accumulates that a warming planet will cause widespread and mostly harmful effects, scientists and policy makers have proposed various mitigation strategies that might reduce the rate of climate change. For those officials in government who must plan now for an uncertain future, however, strategies for adapting to climate change are equally important."[33]

> — *A Survey of Climate Change Adaptation Planning*, The H. John Heinz III Center for Science, Economics and the Environment states:

8

"An equitable international response to climate change must include action on both adaptation and mitigation. Adaptation and mitigation are not choices: substantial climate change is already inevitable over the next 30 years, so some adaptation is essential."

— The Stern Review Team, Report to the Chancellor
of the Exchequer and the Prime Minister[34]

"The people of the world and their governments must find the will and the means to slow, stop and reverse the buildup of global warming gases in the atmosphere to avert catastrophic warming. But it is too late to avert serious consequences, so we must also learn to adapt to a warming world."

— Jonathan Lash, President, World Resources Institute,
"Weathering the Storm"[35]

King County, WA, Executive Ron Sims noted in his introduction to *Preparing for Climate Change: A Guidebook for Local. Regional, and State Governments*, "There was a time, not long ago, when it was not acceptable to talk about adapting to — or preparing for — climate change. The reasoning was that time spent preparing for adapting to the harmful effects of greenhouse gas pollution would divert resources from the essential need to reduce the emission of those gases. . . . there are still many people reluctant to talk about specific adaptation or preparedness policies. But as responsible public leaders, we cannot afford the luxury of not preparing. . . . We must prepare for the impacts underway while we work to avoid even worse future effects."[36]

Or in the words of the IPCC: "There is high confidence that neither adaptation nor mitigation alone can avoid all climate change impacts; however, they can complement each other and together can significantly reduce the risks of climate change."[37]

Local Governments Must Lead

Just as local government officials have taken the lead on mitigation — reducing their local greenhouse gas emissions — they also need to be on the front lines of adaptation — preparing for the local impact of climate change. In the United States, it was Seattle Mayor Greg Nickels who challenged his fellow mayors to pledge to meet or beat the targets for greenhouse gas emissions set by the Kyoto Protocol. The mayors of over 825 cities across America had signed the U.S. Mayor's Climate Protect Agreement by April 2008.[38]

9

Part of what has driven local leaders to step up is the absence of federal leadership on climate change. But it is also the realization that "it is in their jurisdictions that climate change impacts are felt and understood most clearly."[39] As the authors of *Preparing for Climate Change: A Guidebook for Local, Regional, and State Governments*, also noted, "Climate change is a global trend, but one which localities, regions and states will experience to different degrees in different ways . . . Preparing for climate change is not 'one size fits all' process. . . . Preparedness actions will need to be tailored to the circumstances of different communities. . . ."[40]

The experts at the Heinz Center agree on the need for a local, grassroots approach to planning for climate change impacts: " . . . every community is unique in its setting and people, and therefore faces environmental and societal vulnerabilities that will differ from neighboring communities. Understanding the nature of these vulnerabilities is part of the challenge of creating an adaptation strategy."[41]

In fact, according to the World Bank, successful adaptation can only happen if it is driven and implemented at the local level. "Successful adaptation to climate change will require local level institutions that foster collective action on a range of key tasks, such as managing natural resources, mediating competition over scarce resources to prevent insecurity and conflict, mutual aid and community — based infrastructure, development and maintenance."[42]

According to *Preparing for Climate Change*, one of the primary reasons local, regional, and state governments should be proactive in preparing for climate change impacts is: "Planning for the future can benefit the present . . . many projected climate change impacts are in fact more extreme versions of what communities are already experiencing today as a result of present day climate variability and extreme climate events." For example, helping communities prepare for future water shortages and drought by instituting a water conservation/management program will have immediate benefits.[43]

Planning and preparing now can reduce future costs (look at the costs of building a reservoir now, versus in 30 years), reduce future risks, increase future benefits. The authors of *Preparing for Climate Change* explain, "Deferring planning until climate change is 'here' could cause costly delays and increase vulnerability to climate impacts given the time it takes to implement some preparedness strategies. For example, expanding a water supply system to accommodate the combined impacts of population growth and climate change may take 10 to 30 years before the additional capacity is online. The delay could leave a region vulnerable to drought, higher water rates and broader economic costs.

10

"In some cases waiting will foreclose a lower cost preparedness option, leaving . . . only expensive ways out. For example, a low cost strategy for managing the risk of more frequent or intense floods might be to leave a floodplain undeveloped. . . ." But delays in planning may allow development to continue in the floodplain, and the solution later could be more expensive options, including installing dikes or relocating residents.[44]

In its publication *CEO Briefing*, the UNEP notes, "Dangerous climate change is approaching fast. Within 35 years, the costs of climate change could rise to 1 trillion USD in a single year. Adaptation can avoid that scenario, with many other benefits."[45]

The IPCC also notes that climate change preparation provides local government with the opportunity to integrate risk reduction, and readiness for extreme weather, into development planning. "Adaptation measures are seldom undertaken in response to climate change alone but can be integrated within, for example, water resource management, coastal defense and risk-reduction strategies."[46]

According to the UNEP, "Mainstreaming climate change is key. Managing Climate Change should be integrated into policy like water management, disaster preparedness or land-use planning at every level of decision making. The solution is to build local capacity and resilience in a way that links sustainable development, risk management and adaptation for a win-win-win situation."[47]

Daniel Sarewitz, the director of Arizona State University's Consortium for Science, Policy and Outcomes, concludes, " . . . defining adaptation as sustainable development, would allow a focus both on reducing emissions and on the vulnerability of populations to climate variability and change . . ."[48]

Community Leaders Are Equal to the Challenge

Reversing climate change will take decades. In the meantime, it is possible to protect our communities, our economy, and our environment from thoroughly predictable natural disasters through community-based adaptation efforts. Sustained action to reduce or eliminate risk to people and property from hazards and their effects is imperative and well within our ability.

In fact, many communities have already taken action to reduce their risk from natural disasters and climate change. In the last 25 years, flood-, earthquake-, tornado-, and other disaster-prone communities have come together to identify their risks, to identify what measures they can take to reduce the impacts of these risks, and to generate the political, public, and resource support needed to implement these risk-reduction measures. (See Sidebar 1)

11

Sidebar I
Community-Based Hazard Mitigation at Work
Freeport, NY

The Village of Freeport is located on the southern shore of Long Island in Nassau County, New York, approximately 13 miles east of John F. Kennedy Airport. From the start, Freeport relied on its waterfront location; it began as a fishing port and is now the recreational boating center of Long Island. Development in Freeport over the years resulted in frequent flooding, especially in the commercial district known as the Nautical Mile located in an area known as South Freeport.

In 1983 Freeport began to routinely elevate streets in South Freeport. Because of the cost, the time to complete the elevation of all streets at flood risk was estimated to be decades. To this point, the majority of the financing, between $1 and 2 million annually, came from the issuance of general obligation bonds. However, periodically after 1983, Freeport has received financial assistance from both the state and federal Departments of Transportation. By the mid-1990s, many streets had been elevated, including Woodcleft Avenue, which is now a fishing and tourist attraction as well as the most significant commercial business district in Freeport. The Village of Freeport and private citizens raised $10 million to redevelop the Nautical Mile, including the installation of new bulkheads, replacement of overhead electric wires with underground wiring, and construction of new upscale restaurants.

Freeport used funding from six FEMA hazard mitigation grants since 1997 to elevate roads and 23 individual residences. Freeport used grant funds from FEMA's Project Impact to fund public-awareness activities, replace and repair bulkheads, conduct a roadway grade raise and drainage improvement project, remove trees that threatened overhead power lines, and install hurricane-resistant windows and doors in the Village's emergency operations center.

Source: National Institute of Building Sciences (NIBS), "NATURAL HAZARD MITIGATION SAVES: An Independent Study to Assess the Future Savings from Mitigation Activities." http://www.nibs.org/MMC/ MitigationSavingsReport/natural_hazard_mitigation_ saves.htm

Deerfield Beach, FL

Deerfield Beach, Florida, a coastal community of over 66,000 people, was the first Project Impact community to partner with FEMA. Deerfield Beach is well acquainted with damages a natural disaster can cause a community. Having been hit by seven major hurricanes in 75 years, residents knew more hurricanes were statistically almost a certainty. The community's determination to decrease damages sustained from future hurricanes grew after a particularly bad blow from Hurricane Andrew in 1992, followed by near misses of Erin and Opal in 1995.

With guidance from FEMA, Deerfield Beach identified and prioritized mitigation projects that would be most beneficial to the community. One of the first efforts undertaken was retrofitting the Deerfield Beach High School, which also serves as a community shelter during emergencies. Hurricane straps were added to the cafeteria and auditorium, and wind shutters were placed on all the school's windows. Additional community projects included shuttering and disaster-resistant improvements to critical facilities, mentoring, shuttering for single-family residences for senior citizens and low-income households, and a variety of pubic awareness activities.

Deerfield Beach worked very closely with a variety of business partners. The local Home Depot maintained a "Project Impact Aisle," offering products and informational materials on making buildings more disaster-resistant. During the initial two years of the program, the store also designated a senior manager as a Project Impact advocate, allowing him to spend 80 percent of his time in support of Project Impact activities. Solutia, Inc. donated hurricane-resistant glass to retrofit the Deerfield Beach Chamber of Commerce. Deerfield Builders Supply was a corporate sponsor of the annual Hurricane Awareness Week and member of the Local Mitigation Strategy working group, and it donated labor to install windows and doors in the Chamber of Commerce. Marina One Yacht Club built the first hurricane-resistant marine storage facility, designed to withstand 125 mph winds, offering 2,600,000 cubic feet of storage.

Source: FEMA, "Emergency and Risk Management Case Studies Textbook." http://training.fema.gov/EMIWeb/edu/emoutline.asp

continued

Wilmington, NC

Wilmington is a coastal community in North Carolina that was impacted by six hurricanes between 1996 and 1999. Wilmington is one of the original seven Project Impact communities, and the Port of Wilmington, the center of the community's economy, was an active partner in Project Impact.

As part of its commitment to reducing the damage from future hurricanes, the Port of Wilmington invested in a planning and mitigation effort that brought together federal, state, local, and private-sector organizations. A risk analysis and four mitigation measures were implemented, designed to eliminate or minimize the hurricane loses and ensure business continuity. These measures include securing of gantry cranes and mobile cranes to ensure they do not topple over; nonstructural measures to secure sensitive equipment; structural reinforcement of buildings, including wind resistant roofing; and structural reinforcement of electric power and telecommunications systems.[1]

A study of the benefits of Project Impact to the Wilmington's labor market concluded, "The findings of this article are consistent with Project Impact's having a beneficial impact on the labor market of Wilmington. After the initiative, the equilibrium unemployment rate in Wilmington is significantly lower than before the policy intervention, controlling for the effects of other business cycle factors and trends. Additionally, the policy intervention is associated with a significant reduction in the long-run variance of the unemployment rate. The evidence is also consistent with the claim that the disturbance created by a hurricane is less after Project Impact was initiated than before. Taken together, these findings suggest that the activities and coordination efforts associated with Project Impact coincide with improvements in the Wilmington labor market characterized by a lower natural unemployment rate and a reduction of labor market risk. On one hand, these findings may be taken as evidence that Project Impact can improve the performance of a local economy. The results suggest that, at the very least, increased interaction between public and private sectors may be associated with improved labor market conditions."[2]

Sources: 1. North Carolina Division of Emergency Management, "Sustainable Infrastructure and Critical facilities," in *Mitigation in North Carolina: Measuring Success* (2002). 2. Bradley T. Ewing and Jamie Brown Kruse, "The Impact of Project Impact on the Wilmington, North Carolina, Labor Market," *Public Finance Review* (2002): 30; 296. http://pfr.sagepub.com/cgi/content/abstract/30/4/296

In the following chapters, we present essays and case studies that illustrate how community leaders around the country have successfully addressed their communities' hazard risks. We believe that their example will serve as a guide for community leaders around the country in addressing the heightened hazard risks caused by global warming.

NOTES

1. IPCC. Climate Change 2007: Synthesis Report, *Summary for Policymakers: An Assessment of the Intergovernmental Panel on Climate Change,* November 2007.
2. Intergovernmental Panel on Climate Change, www.ipcc.ch/about/index.html.
3. TheNobelPrize,2007,http://nobelprize.org/nobel-prizes/peace/laureates/2007/.
4. *New York Times,* "Science Panel Calls Global Warming 'Unequivocal,'" February 3, 2007, nytimes.com/2007/02/03/science/earth/03climate.html.
5. IPCC, 2007, Summary for Policymakers, *Climate Change 2007: Impacts, Adaptation and vulnerability,* Contribution of Working Group II to the Fourth Assessment of the IPCC (April 2007).
6. IPCC. Climate Change 2007: Synthesis Report, *Summary for Policymakers: An Assessment of the Intergovernmental Panel on Climate Change,* November 2007.
7. Ibid.
8. Union of Concerned Scientists, "Findings of the IPCC Fourth Assessment: Climate Change Science," www.ucsusa.org/global_warming/science/ipcc-highlights.html.
9. Ibid.
10. Ibid.
11. IPCC. Climate Change 2007: Synthesis Report, *Summary for Policymakers: An Assessment of the Intergovernmental Panel on Climate Change,* November 2007.
12. Union of Concerned Scientists. "Findings of the IPCC Fourth Assessment: Climate Change Science," www.ucsusa.org/global_warming/science/ipcc-highlights.html.
13. Ibid.
14. IPCC. Climate Change 2007: Synthesis Report, *Summary for Policymakers: An Assessment of the Intergovernmental Panel on Climate Change,* November 2007.
15. Union of Concerned Scientists, "Findings of the IPCC Fourth Assessment: Climate Change Science," www.ucsusa.org/global_warming/science/ipcc-highlights.html.
16. Ibid.
17. IPCC. Climate Change 2007: Synthesis Report, *Summary for Policymakers: An Assessment of the Intergovernmental Panel on Climate Change,* November 2007.

18. Union of Concerned Scientists, "Findings of the IPCC Fourth Assessment: Climate Change Science," www.ucsusa.org/global_warming/science/ipcc-highlights.html.

19. Ibid.

20. IPCC, *Climate Change 2007: Impacts, Adaptation and Vulnerability,* Contribution of Working Group II to the Fourth Assessment of the IPCC, April 2007.

21. Ibid.

22. IPCC. Climate Change 2007: Synthesis Report, *Summary for Policymakers: An Assessment of the Intergovernmental Panel on Climate Change,* November 2007.

23. ICLEI — Local Governments for Sustainability, Climate Change Impacts Group, King County Executive Office, "Preparing for Climate Change: A Guidebook for Local, Regional, and State Governments," September 2007.

24. World Bank Group. Environment Matters Annual Review, "Climate Change and Disaster Risk Reduction," Maarten Van Aaist, June 2006–July 2007.

25. Ibid.

26. Ibid.

27. *New York Times,* "Science Panel Calls Global Warming 'Unequivocal,'" February 3, 2007, nytimes.com/2007/02/03/science/earth/03climate.html.

28. Ibid.

29. Judith Rodin, President of the Rockefeller Foundation, "Climate Change Adaptation: The Next Great Challenge for the Developing World," remarks to the American Association for the Advancement of Science, February 15, 2008.

30. Science Daily News release, "Adaptation to Global Climate Change Is an Essential Response to a Warming Planet," February 8, 2008.

31. Ibid.

32. IPCC, *Climate Change 2007: Impacts, Adaptation and Vulnerability,* Contribution of Working Group II to the Fourth Assessment of the IPCC, April 2007.

33. The H. John Heinz III Center for Science, Economics, and the Environment, "A Survey of Climate Change Adaptation Planning, October 2007.

34. Stern Review Team, "What is the Economics of Climate Change?" discussion paper 31 (January), report to the Chancellor of the Exchequer and the Prime Minister.

35. World Resources Institute, "Weathering the Storm: Options for Framing Adaptation and Development," 2007.

36. ICLEI — Local Governments for Sustainability, Climate Change Impacts Group, King County Executive Office, "Preparing for Climate Change: A Guidebook for Local, Regional, and State Governments," September 2007.

37. IPCC, *Climate Change 2007: Impacts, Adaptation and vulnerability,* Contribution of Working Group II to the Fourth Assessment of the IPCC, April 2007.

38. The U.S. Conference of Mayors, Climate Change Center, http://usmayors.org/climateprotection/climatechange/asp.

39. ICLEI — Local Governments for Sustainability, Climate Change Impacts Group, King County Executive Office, "Preparing for Climate Change: A Guidebook for Local, Regional, and State Governments," September 2007.

40. Ibid.

41. The H. John Heinz III center for Science, Economics and the Environment, "A Survey of Climate Change Adaptation Planning, October 2007.
42. World Bank Group, *Environment Matters Annual Report,* June 2006–July 2007.
43. ICLEI — Local Governments for Sustainability, Climate Change Impacts Group, King County Executive Office, "Preparing for Climate Change: A Guidebook for Local, Regional, and State Governments," September 2007.
44. Ibid.
45. CEO Briefing: Innovative Financing for Sustainability, United Nations Environmental Programme (UNEP), November 2006.
46. IPCC, 2007 Summary of Policymakers, *Climate Change 2007: Impacts, Adaptation and vulnerability,* Contribution of Working Group II to the Fourth Assessment of the IPCC, April 2007.
47. CEO Briefing: Innovative Financing for Sustainability, United Nations Environmental Programme (UNEP), November 2006.
48. Science Daily News release, "Adaptation to Global Climate Change Is an Essential Response to a Warming Planet," February 8, 2008.

2

Planning and Protecting the Environment

INTRODUCTION

In order to effectively address the increased disaster activity caused by global warming, it will be necessary to involve new stakeholders in the process and to consider a wider array of risk-reduction measures that can be employed to reduce future disaster impacts. The two essays presented in this chapter directly address the need to expand the universe of skills and actions to be successful in addressing global warming.

The first essay examines how urban and regional planners can play a productive role in the design and creation of hazard-mitigation programs designed to reduce the impacts of future disasters aggravated by global warming. As Jim Schwab notes in his essay, "Planning must begin to incorporate a full suite of options, both for reducing greenhouse emissions in order to slow the process of climate range and a more sophisticated assessment of options for local and regional hazard mitigation (including incorporating the best mitigation available in postdisaster recovery and reconstruction) in order to achieve a truly sustainable society and economy."

The second essay describes the impact unwise environmental policies have had in damaging and destroying wetlands and forests that provide natural protection to communities from floods and wildfires. Author Kathryn Hohmann examines the series of missteps that have put our communities and citizens at greater risk and presents potential solutions in the form of "collaborative projects — networks operating at the grassroots level to create change. Drawing on the strengths of diverse disciplines, these partnerships help communities manage risk; they preserve environmental quality; they lower costs of emergency response. If these networks succeed, they may expand the constituency for positive change, renew our communities, and literally save our world."

We hope that these two essays will provide some guidance to community leaders in terms of what the planning community can bring to building partnerships and policy guidance for addressing the impacts of global warming on community risks and the important role a healthy, natural environment means to the safety of a community and its residents.

THE ROLE OF PLANNING IN REDUCING IMPACTS OF GLOBAL WARMING

Jim Schwab, AICP

Jim Schwab, AICP is senior research associate at the American Planning Association (APA). Mr. Schwab served as the primary author and principal investigator for *Planning for Post-Disaster Recovery and Redevelopment* (PAS Report No. 483/484, December 1998), which APA produced under a cooperative agreement with the Federal Emergency Management Agency. He served as the project manager for a FEMA-supported project in which APA developed training for planners on the planning provisions of the Disaster Mitigation Act of 2000, and for the Firewise Communities Post-Workshop Assessment, a contract with the National Fire Protection Association to determine the impact of its Firewise workshops on community behavior. He is currently managing a project on Planning for Urban and Community Forestry underwritten in part by the USDA Forest Service. Mr. Schwab earned one M.A. in journalism and another in urban and regional planning from the University of Iowa, and earned a B.A. in political science at Cleveland State University.

Introduction

Planning at the community and regional level is essentially the practice of developing a vision for that community's future welfare and development, embodying that vision in an adopted plan, and implementing that vision through regulations, incentives, public policies, and administrative actions. Part of that process involves issues of land use, but other parts relate to economic development, public investment, infrastructure, and environmental policy, as well as other tools and techniques that may be available for the community to accomplish its stated goals. The exact form and focus of a community's plan will often depend on the statutory authorities it has to pursue those goals, as well as the authorities delegated to neighboring, superior, or special jurisdictions that may cooperate with or hinder the vision behind the plans. For instance, a community's goal of compact growth may rely to a significant extent on the cooperation of an independent school district in its school siting policies, or on a regional water board or transit authority in locating infrastructure or developing new bus or rail lines. The challenge is not necessarily that the various jurisdictions want to obstruct change, but that each has its own institutional dynamics.

That various local political systems evolved without reference to the problem of climate change is inherently obvious. They have not always evolved even in reference to existing and widely recognized public policy challenges, such as air pollution, groundwater protection, or regional transportation efficiency. Planning thus faces a serious challenge almost everywhere in the United States in finding ways to coordinate meaningful responses to the need to reduce the potential impacts of climate change. Some of the methods of mitigating those impacts fall directly within the realm of one local government or another; others may require a good deal of political persuasion and regional cooperation in order to make a difference. This essay is a discussion of the most important ways in which planning may be able to make those changes.

Framework for Reducing Impacts

In most communities comprehensive planning has become a linchpin of the entire planning process. Ideally, by encouraging broad public participation in developing a vision of what they would like to see in their community's future, then building a plan around that vision that addresses its many facets through elements dealing with such essential issues as land use, transportation, housing, and economic development, planners can help decision makers focus on the most important policy choices for achieving that vision and see the relationships between those issues. For instance, the most highly discussed relationship in the planning profession is probably that between transportation and land use, because the two are so highly dependent on each other.

When assessing the potential risk from climate change, community planners and elected officials face an almost bewildering array of potential influences and impacts, with relatively few historical precedents for incorporating them into comprehensive plans. Most communities have only

21

recently begun to consider these risks, if they have done so at all, yet planning urgently needs to establish a viable framework for doing exactly that. As of February 2008, only one state had passed new legislation directing local and regional planners to incorporate provisions for addressing climate change in comprehensive, regional land use and transportation plans. Unsurprisingly, that state was California, which was also locked in a legal battle with the U.S. Environmental Protection Agency over its right to a waiver from federal Clean Air Act regulations to allow stricter regulations of its own concerning greenhouse gas emissions.[1]

The California approach, which one hopes may become more common but whose goals can be pursued by local and regional planning bodies elsewhere even without state mandates, engenders some other critical questions regarding the efficacy of such planning. For one, it would help enormously if state agencies with special technical resources would not only share critical climate change data with local and regional planners but also provide planning grants to support those efforts and develop measurement tools to ease the burdens on local planners to acquire the necessary expertise to incorporate the appropriate data into their plans. Local planners will need considerably better information in many instances in order to fully understand the options available to them in the form of policy changes, practical public investment options, and conservation measures for maximizing the effectiveness of public efforts to reduce greenhouse gas emissions. These can include higher density around mass transit, energy-efficient affordable housing, water conservation, and even some major shifts in energy distribution systems. The list of options is limited largely by our collective civic imagination and technical creativity.

These same caveats apply equally well to the assessment of natural hazards risks such as those described below. Planners have particular expertise in urban design and the planning process but are for the most part generalists when it comes to many scientific and technical subjects. Their job is not to be experts on climate change but to integrate such information with other aspects of planning where relevant, particularly issues like transportation and local energy policy. In fact, the California Chapter of the American Planning Association in late 2007 produced policy principles advancing recommendations for adopting many of these ideas.[2]

Framework for Assessing Risk

Another way to address these risks is through a natural hazards element in the comprehensive plan, which can incorporate assessments of climate

22

change impacts and provide a means for identifying specific actions and recommendations in other plan elements for dealing with those problems. So far, only about ten states nationwide require some sort of hazards-related element in local comprehensive plans; most states still do not require a comprehensive plan in the first place, but most at least specify or suggest the kinds of elements that should be included in such a plan in order for it to be complete.[3]

However, comprehensive planning is only one route for developing such an assessment, even though it potentially guarantees the best opportunity for relating natural hazards to other aspects of community planning. Communities have long developed various specialized plans in response to state and particularly federal funding incentives for housing, environmental protection, and, most recently, natural hazards. In 2000, Congress passed the Disaster Mitigation Act (DMA), which requires the adoption of a local hazard-mitigation plan approved by the Federal Emergency Management Agency (FEMA) in order to be eligible for either competitive Pre-Disaster Mitigation (PDM) grants or postdisaster Hazard Mitigation Grant Program (HMGP) funds.

Some noteworthy obstacles have emerged to widespread integration of such planning into the daily routine of planning activities at the state and local levels. The biggest is that both FEMA itself and the state agencies that handle emergency management duties and relate to FEMA in the national chain of command have grown up in a culture built originally around civil defense and, subsequently, around emergency response. Historically, only a modicum of interdisciplinary communication has taken place between these people and urban and regional planners, who have concerned themselves more with land use, urban design, transportation, and economic development, and only occasionally, or under the pressure of unexpected events, with natural hazards. Thus, even when the passage of the Disaster Mitigation Act established powerful incentives for preparing and adopting state and local hazard mitigation plans, the task inevitably fell in most cases to emergency managers, and the same pattern of noncommunication with planners continued, in spite of the implications that such plans had for the long-term future of communities.

Most local plans, therefore, are still being prepared by emergency managers with at best tangential involvement by local planners, who themselves often fail to assert a role for themselves in this process. There is also no direct remedy for this problem at the federal level, because it is not the place of the federal government in our federal system to determine who in local government should undertake the task of preparing the plan. The

23

FIGURE 2.1 Caruthersville, MO, 4-7-06. — Janet L. Sanders, a building and planning superintendent for the City of Jackson, MO, works with a Caruthersville, MO, city map to track the houses that have been inspected for damage. She and other volunteer engineers are working with FEMA employees to provide assistance and information to the townspeople of Caruthersville, MO, affected by a tornado that hit the town. Photo by Patsy Lynch/FEMA.

best the federal government can do — and probably will do with increasing efficacy over time as the regulations for DMA evolve — is to score compliance on the basis of how the lead agency preparing the local plan solicits input from other agencies in order to ensure broad input and participation. That said, it is often one thing to check a box saying people participated and quite another to have structured a process in which such participation was valued and of high quality. It is fair to say that planners are generally better trained for organizing such input than most other professionals in local government. That makes it critical that, as the impacts of climate change compound the effects of existing natural hazards, planners assert for themselves a larger role in the process and marshal all appropriate expertise for ensuring the adequacy of hazard-mitigation planning.

There are exceptions to this general pattern. As previously noted, some states require the inclusion of some form of hazards element in the local comprehensive plan. Florida, for instance, has detailed legislation describing the proper content of a comprehensive plan element addressing coastal hazards. Florida has plenty of reason to be concerned about this topic, but so do Alabama, Mississippi, Louisiana, and Texas, none of which have seen fit to enact similar mandates. Oregon requires that all local plans address state planning goals, one of which concerns hazards, and the state also provides significant technical guidance on the subject to advance the process. California requires a safety element that originally focused on seismic hazards but has been expanded over the years, most recently to include floodplain management. In these cases, the mechanism is in place for forcing the issue, and FEMA is already working to make a thoroughly crafted hazards element an adequate basis for presumptive compliance with DMA.

The reason all this matters for the role of planning in addressing climate change is that implementation of effective responses to climate change and the hazards it may exacerbate depend on an effective assessment of those impacts and a sound understanding of how those impacts relate to and affect each other. The rest of this chapter is devoted to exploring three major manifestations of hazards related to climate change and how planning may be able to respond to them.

Extreme Heat Events

One of the most frequently cited impacts of global warming is an increase in average temperatures, particularly at higher latitudes. In addition to higher average temperatures, these scenarios typically forecast growing numbers of days in various cities of temperatures above certain high levels, for instance, days above either 90° or 100°F. When these days occur over a period of days rather than sporadically, they become known as extreme heat events, or in popular parlance, heat waves. In many major cities, they have been associated with major threats to public health as elderly and other vulnerable population groups succumb to heat and humidity, becoming ill or dying. One particularly noteworthy case, the heat wave that afflicted Chicago in 1995, became the focus of a book by sociologist Eric Klinenberg that examined the social characteristics of the victims. Its finding was that the more than 500 deaths that occurred resulted not merely from age or disability but also from conditions of social isolation that allowed the victims to die without being noticed until it was too late.[4]

What happened in Chicago that summer was tragic but not unusual. A 2007 study by the Center for Integrative Environmental Research at the University of Maryland provides comparative estimates of average heat-related mortality for average summers in a variety of cities across the United States, with Chicago's present total of 173 being second to New York, a much larger city, but Philadelphia, Detroit, and St. Louis all top 100. The chart then compares those numbers to five estimated averages under three different climate change scenarios, with most numbers climbing significantly.[5]

25

Most of Klinenberg's *Heat Wave* is devoted to the failures of social service delivery that allowed victims to die in isolation, and to documenting his key finding that deaths occurred disproportionately in neighborhoods lacking social cohesion. This point was buttressed by his demonstration that the Little Village neighborhood, a relatively poor but socially cohesive Hispanic neighborhood, suffered far fewer deaths than others with equivalent socioeconomic status. At first blush, this may seem to be a problem primarily for social workers and public health officials that has little to do with planning. Since the problem has drawn relatively little attention in planning literature, one might also assume that planners themselves share that perception, and that assumption is probably accurate.

Such an assumption would be unfortunate, however, and may become more unfortunate if the impacts of climate change exacerbate the public health threats of future urban heat waves. Issues of urban design, community development, transportation, and even of the applications of geographic information systems (GIS) have everything to do with how many lives may ultimately be saved or lost in future heat emergencies. Planners, who are already engaging with questions of how urban design affects physical fitness by encouraging or discouraging people from walking instead of driving to their daily destinations, can certainly reexamine how zoning and other public policies may facilitate or discourage the interaction of neighbors in inner-city communities that have typically suffered high casualties as a result of extreme weather. One factor affecting the willingness of vulnerable elderly citizens to venture beyond their apartments, for instance, is the prevalence or absence of crime and the overall social atmosphere of the neighborhood. Employing the growing range of studies about how everything from trees to the placement of social service facilities affects these attitudes, planners can help to ensure that, as neighborhoods evolve and redevelop over time, their pattern becomes more humane and inviting of human interaction.

If this focus seems remote from the concerns of a planetary issue like climate change, it is not. It is part of our best bet, along with concerted efforts to mitigate urban heat islands, for making our cities livable in spite of larger changes that lie beyond the control of individual municipalities. In fact, regaining some control over the livability of urban neighborhoods may even inspire some sense of empowerment among citizens that they can still stem the larger tide of climate change. Combining this effort to reduce the death toll with larger efforts to reduce the consumption of energy through conservation and efficiency can buy all of us valuable time to address the larger global issues at hand.

In comparison to Chicago in 1995, the improved capabilities of GIS would suggest that planners, social workers, and public health officials can collaborate effectively to map the precise locations of the most vulnerable populations so that, when heat emergencies arise, the latter two groups are well prepared to reach out in a timely and effective manner to extend needed relief or shelter to otherwise isolated individuals at high risk of dying. This would be an excellent integration of planning and social service capacities within city government.

It is also important to recognize how planning can make the entire public as well as its decision makers aware of the interrelationship between such dire issues as urban heat emergencies and the need for progressive environmental planning. The role of the urban forest, for example, remains inadequately understood among both planners and elected and appointed officials. Trees serve numerous purposes in mitigating the urban heat island, including reducing summer cooling needs (and resultant energy use) by double-digit percentages and providing an outdoor respite from the sun. This is all in addition to other environmental benefits such as filtering air pollution and reducing storm water runoff. Adding in the global issue of carbon sequestration, it becomes clear that planning for the health of the urban forest is not an aesthetic priority so much as a high priority for preserving urban environment and quality of life, particularly in the face of new challenges from climate change.

Finally, although the equation is not always tightly drawn, it is critical to understand that the larger metropolitan picture of compact development versus sprawl, and how compact development can facilitate the economic viability of mass transit as a way of reducing vehicle miles traveled and fuel consumption, are vital elements influenced by planners and in turn influencing the release of the greenhouse gases that result in global warming.

Coastal Hazards

Sea-level rise may well be the single issue most prominently associated in the public mind with global warming. The problem it poses for planners in motivating a response from citizens and public officials alike is that it is a slow process that is difficult to see from year to year. It is not readily apparent to most people how a rise of a few inches actually affects the viability of development along shorelines that often represent powerful attractions. The fact that more than half of the U.S. population lives in the nation's 673 coastal counties[6] underscores the challenges planners may face in explaining the need for greater caution in building near the water as a result of climate change over the next century. For many people, the attraction remains immediate while the threat is abstract and remote. The attraction often remains even in spite of dire catastrophes like Hurricanes Katrina and Rita.

From a planning perspective, the best practices in mitigation of coastal hazards are already well established, albeit far from universally practiced. Political pressures to allow beachfront development are often powerful, in large part because such property is among the most valuable real estate in the nation. Until market forces finally begin to incorporate a significant discount related to the high risks involved, beachfront property in coastal storm zones will continue to be densely developed. Market forces have been blunted in this respect by expectations of federal assistance and the availability of flood insurance, although there are signs that this distortion may begin to reverse itself. In the meantime, local officials often find it hard to "just say no" to questionable or unwise proposals, in part because, where planning is weak, they may not even be well informed about the level of risk such development entails. Many of the communities in southern Mississippi, for instance, not only did not have planning staff; they also lacked planning commissions and even zoning ordinances. In such circumstances, it is hard to say that decisions about locating new development had a solid foundation.

Existing best practices, however, can go a long way in protecting most communities from the worst impacts of major hurricanes. These include setbacks such as those provided by Florida's Coastal Construction Control Line, which provides a way of calculating long-term coastal erosion and requiring that development not occur seaward of that line. Although Florida could easily become more restrictive — for instance, by adopting a stricter definition of a high-hazard zone than its current statutory standard of a Category 1 evacuation zone — drawing such lines in the sand at least helps to prevent some of the more egregious possibilities, with hotels and homes built square to the ground, awaiting the full force of a 20-foot or larger storm surge. Elevating homes and other buildings within hazard zones has also preserved numerous structures and lives. Adopting standards like the South Florida Building Code to ensure hurricane-resistant construction — and then actually enforcing them — has also been effective. And the principles of subdivision and community design, orienting evacuation routes away from the coast and locating buildings further inland, plus preserving intact dune systems, are all part of the planner's tool box. These techniques are readily available to any community-planning department willing to do the research, and they will be more important if certain expected impacts of climate change manifest themselves. In many respects, it may be more important in mitigating impacts of climate change in areas subject to coastal storms that communities use the methods already widely known but not yet widely adopted than to worry about what new things planning can do to ameliorate the situation.

Nonetheless, sea-level rise may well be an imminent reality that can make today's high-hazard zone maps obsolete over time. The best answer may be to do more of the same, and to do it more intensely. If a high-hazard line drawn along the coast prohibits seaward development now, factor into the calculation of that risk the likely values of sea-level rise based on the best data available, and draw a new line farther inland to account for the fact that the encroachment of the water's edge will wipe out, at least over the life of many new buildings, much of the beachfront real estate that was supposed to provide a buffer from nature's rage when the waters rise. What is the necessary safety factor required by sea-level rise? Add it in, and move the high-hazard line accordingly. As each new storm damages some buildings beyond 50 percent of fair market value, buy them out and retire them from the scene as nonconforming uses.[7] Over time, it may become possible to reduce the high levels of exposure that still afflict many Atlantic coast and Gulf Coast communities.

Wildfires

One of the central principles in climate change is that changes are not uniform around the world. There can be vast differentials in impact among desert regions, polar regions, temperate zones, and coastal and inland areas, and at different latitudes. One of the most significant alterations projected is a shift in precipitation patterns accompanying changes in average seasonal temperatures. Some areas may experience more extreme droughts while others may receive considerably more rain, depending on how existing atmospheric patterns and ocean currents are disrupted.[8]

One of the more problematic possibilities is that of more extreme drought in some existing arid areas, combined with more extreme heat. While these two impacts constitute hazards in and of themselves, together they also constitute a prescription for increased risk of wildfire. In the western United States, in particular, this prospect exacerbates an existing problem of high fuel buildup in areas where public policy through most

29

of the 20th century demanded aggressive fire suppression. The resulting accumulation of flammable underbrush and dense stands of small trees in areas that historically had been naturally much more thinly forested has produced more severe wildfires than would have occurred otherwise. Add to this troubling equation rapid population growth in the West, coupled with the attraction for many people of locating primary residences or vacation homes on forested hillsides in rural communities.[9] The outcome is that the very presence of many of these homes in the wildland-urban interface often represents an obstacle to the use of such proven wildfire mitigation techniques as prescribed burns (to reduce the likelihood of major wildfires) while tending to obligate local fire departments to rescue such homes and their residents when fires occur. Fire-response policy thus tends to be distorted by this new human presence.

Again, as with coastal hazard zones, the best planning practices in this area are largely already known, even though we are learning more over time. The Firewise Communities training programs have taught thousands of public officials, planners, architects, home owners association leaders, and businesspeople the principles of wildfire mitigation for nearly two decades. These include defensible space, which primarily involves keeping flammable vegetation away from structures; ensuring that new developments have more than one access route so that residents are not trapped by fire; adopting tough fire-resistant building codes, such as the amendments that California approved in 2005, which take effect in 2008,[10] and simply not allowing development in some areas where its presence will either unduly complicate firefighting efforts or pose unacceptable risks, particularly on steep hillsides or in other especially rough terrain. As with areas facing hurricane hazards, the ability of local officials to say NO may often be as important as the ability to craft tougher rules. For one thing, development in many wildfire-hazard zones is especially likely to promote sprawl, when many western cities still have room for much safer infill development.

At the same time, as in many coastal areas, what is known about best practices is not always practiced. Because of the pace of development, these jurisdictions are in a race against time, whether they know it or not. The race is to adopt and enforce needed regulations before too much new development adopted without effective standards has already been put in place. A strong orientation toward property rights prevails in much of the West, although the political landscape in this regard is changing as the demographics of the West change due to in-migration, particularly of newcomers who are more environmentally minded and have a stronger devotion to preserving the landscape they have come to cherish. These changes may over time afford planners more opportunities to sell communities on more aggressive efforts at wildfire mitigation. Regulations are only part of the package, though a powerful one.

Final Observations

American society is evidencing a growing interest in green communities and green building design, and various cities are racing to claim the mantle of "green city," whether or not they have taken a sufficiently holistic approach to planning to have earned it. Interest in green communities and buildings is growing at least in part because of growing concern about the impacts of human activities on climate change and a desire to reduce those impacts. However, public understanding of the full range of planning and policy changes needed to make a significant difference is still lagging, and the need for significant change is increasingly urgent. In addition, not only the public but most elected officials and planners lack a fully developed understanding of the full range of impacts of climate change on the hazards their communities will face in coming decades, in part because planning for such hazards is in many ways still so nascent in the vast majority of jurisdictions. A good starting point, however, is for the public, its planners, and elected and appointed officials to understand that there is almost nothing less green and more wasteful than the massive devastation that followed Hurricanes Katrina and Rita. Planning must begin to incorporate a full suite of options both for reducing greenhouse emissions in order to slow the process of climate range, and for a more sophisticated assessment of options for local and regional hazard mitigation (including incorporating the best mitigation available in postdisaster recovery and reconstruction) in order to achieve a truly sustainable society and economy. This is a challenge that should keep all of us busy for decades to come.

Notes

1. That law was A.B. 32, which seeks to achieve carbon reduction goals; for a discussion of the planning implications and planning policy principles, see California Chapter—American Planning Association, "Planning Policy Principles for Climate Change Response," September 2007. Available at: http://www.calapa.org/attachments/contentmanagers/711/ClimateChange.pdf.

U.S. EPA Administrator Stephen Johnson issued the denial of California's waiver in December 2007 lawsuit, and it immediately became both political fodder in the presidential campaign and the target of a California was joined by several other states. See the EPA decision and related papers at http://www.epa.gov/otaq/ca-waiver.htm.

2. California Chapter-APA, pp. 4–5.

3. See "Summary of State Land Use Planning Laws," an annually updated web report produced for the Institute for Business and Home Safety by the American Planning Association, at http://www.ibhs.org/publications/view.asp?id=302.

4. Klinenberg, Eric. *Heat Wave: A Social Autopsy of Disaster in Chicago* (Chicago: University of Chicago Press, 2003).

5. The U.S. Economic Impacts of Climate Change and the Costs of Inaction (Center for Integrative Environmental Research, University of Maryland, October 2007), p. 13. Available online at www.cier.umd.edu/climateadaptation/.

6. Crossett, Kristin, et al., *Population Trends Along the Coastal United States: 1980–2008* (Washington, D.C.: National Oceanic and Atmospheric Administration, 2004), p. 5.

7. In zoning law, a nonconforming use is one that presumably was legal at the time it was built but no longer complies with current zoning restrictions due to subsequent amendments of the code. These uses are allowed to continue, but if destroyed or damaged beyond a certain level, typically 50 percent of value, may not be rebuilt in their existing form. Changes or expansions of existing nonconforming uses typically also trigger imposition of the newer code provisions. The National Flood Insurance Program also disallows rebuilding on the same basis. Some communities go beyond these requirements with stricter limitations, such as applying cumulative damages over a period of time, rather than simply looking at percentage of value with each new disaster.

8. Recent discussion among both climate change experts and many of the journalists who cover environmental issues has focused to some degree on whether the terms "climate change" or "global warming" are even adequate to the task of conveying to the public the real nature of the issue it faces. While both terms have the convenience of having gained a certain familiarity, some are advocating the use of other terms, such as "climate disruption," as more conceptually accurate.

9. For a substantial examination of the demographic forces at work, see James Schwab and Stuart Meck, *Planning for Wildfires* (Chicago: American Planning Association, 2005), Planning Advisory Service Report No. 529/530. Of note is that the five states with the highest percentage increases in population in the 2000 census were all in the Rocky Mountain West.

10. For a summary of the new codes see http://www.fire.ca.gov/fire_protection/fire_protection_prevention_planning_wildland.php.

SIX DEGREES OF SEPARATION: NETWORKS TO PRESERVE WILD PLACES, MITIGATE DISASTER, AND COMBAT CLIMATE CHANGE

Kathryn Hohmann

Kathryn Hohmann holds a Bachelor of Science degree from the University of Minnesota. From 1984 to 1987, she worked as a writer for the U.S. Fish and Wildlife Service, specializing in wetlands conservation and coordinating communications for an international wetlands program that spanned the U.S., Canada and Mexico. Hohmann joined the Sierra Club in Washington, D.C., where she worked from 1990 to 2000. She served as Director of the Sierra Club's Environmental Quality Program, lobbying Congress and the Administration on pollution and land-use issues. She led the Sierra Club's efforts to overhaul the Corps of Engineers nationwide wetlands permit program. Along with David Conrad of the National Wildlife Federation, Hohmann is recipient of the Federal Emergency Management Agency's Public Service Award for her work on preserving the nation's wetlands and protecting communities from floods. In 2000, she moved to Montana, continuing to work at the grassroots level for the Sierra Club, concentrating on federal land-use issues until early 2007. Hohmann currently serves as Southwestern Montana Regional Director for the American Red Cross, coordinating emergency response in thirteen Montana counties.

Imagine that you're a doctor, running a family practice in a rural area. Your patient, an active male in his fifties, seems to have a mild case of the flu — until his fever spikes without warning. When his temperature soars over 104, his breathing turns shallow and rapid and he becomes delirious. You need to act now. Your medical training tells you that the difference between normal body temperature — 98.6 degrees — and a fever that could precipitate seizures, kidney failure, or even a heart attack is little more than six degrees.

There's an equally slim margin between a healthy climate and an ecological emergency. In fact, the "worst-case scenario," developed by scientists on the International Panel on Climate Change describes a climate that's 6.4 degrees (11.5°F) hotter by the end of the century. A difference of 6 degrees might not sound like much, but for a perspective, over the past century, the earth has warmed only 0.6 degrees. Like the hypothetical patient battling a life-threatening fever, under the worst-case scenario, our world would be in critical condition, with parched lands turning to desert, oceans becoming too acidic for marine life, plants and animals dying in waves of mass extinction.

And even in best-case scenarios, scientists predict more frequent, more devastating disasters like hurricanes, killer heat waves, and tornadoes.

Disasters may force action on climate change before species go extinct or wild areas vanish. Of course, it is vital to respond to the human needs that come in the wake of a flood or a wildfire, but it is also critical to tackle climate change with the integrity of the natural world in mind. In order to succeed, climate change policies must explicitly address nature — in part because wild places and our fellow species have an intrinsic right to exist, but also because certain natural systems, like the immune system of a healthy body, can play a critical role in preventing and mitigating disasters. The most intelligent approaches to climate change will necessarily incorporate sound environmental policies that recognize the power of these wild places as stabilizing and risk-reducing forces. Rather than allowing politics and development pressures to trump environmental protection, we must get nature "back on our side." Only if we preserve and restore these ecosystems can we rely on the resiliency of the earth for the long battle against climate change that lies ahead.

This essay concentrates on wildfires and floods, two kinds of natural disasters that are becoming more frequent and severe as the earth warms. The chapter shows how shortsighted environmental policies have contributed to the destruction of wild forests and wetlands, compromising the "disaster-proof" capacities of these two ecosystems, leaving the natural world impoverished and communities at risk. The essay deals with development trends that have left more people in harm's way. Whenever possible, I will rely on personal experience gained as a communications professional for the U.S. Fish and Wildlife Service's wetlands division, as a lobbyist for an environmental organization in Washington, D.C., and serving in disaster response in southwestern Montana, where I currently reside.

After this overview, the chapter turns to solutions and discusses collaborative projects — networks operating at the grassroots level to create change. Drawing on the strengths of diverse disciplines, these partnerships help communities manage risk; they preserve environmental quality; they lower costs of emergency response. If these networks succeed, they may expand the constituency for positive change, renew our communities, and literally save our world.

Smoke Signals Spell out Climate Change

The wildfires that swept the West in the summer of 2002 caused the death of 21 firefighters, drove tens of thousands of people from their homes, and destroyed more than 2,000 structures. More than 7.2 million acres burned across the western United States.

The power of the blazes shocked even veteran firefighters, who witnessed 500-foot-high walls of flames, fire-created tornadoes, fires that could leap highways and burn down and climb steep hills. Something extraordinary seemed to be going on, creating infernos that burned hotter and longer than ever before. Now, groundbreaking research fingers climate change as the culprit. When it comes to wildfires, it seems that climate change is literally fanning the flames.

Dr. Anthony Westerling and colleagues at the Scripps Institution of Oceanography at the University of California, San Diego, researched records of large wildfires, analyzing the severity and duration of fires, and identified striking trends. Since 1986, longer, warmer summers have resulted in a fourfold increase of major wildfires and a more than sixfold increase in the area of forest burned, compared to the years from 1970 to 1986. The length of the average wildfire season has also increased by 78 days, with the Rocky Mountains experiencing most of the climate-influenced wildfires, followed by northern California. They found that the average burn duration of large fires has increased from 7.5 to 37.1 days. Westerling's research, published in *Science* in 2006, suggests that as the climate warms, hotter and drier summers provide tinderbox conditions for wildfires. Although many factors have likely played a role in the number and severity of the fires, "increased temperature has really been driving the increase," said Westerling.[1]

Milder winters also stoke the fires. On the drought-stricken forests of the West, snowpack is like a cool compress. In *Science*, Steven Running writes, "The hydrology of the western United States is dominated by snow; 75 percent of annual stream-flow comes from snowpack. Snowpacks keep fire danger low in these arid forests until the spring melt period ends. Once snowmelt is complete, the forests can become combustible within 1 month because of low humidities and sparse summer rainfall. Most wildfires in the western United States are caused by lightning and human carelessness, and therefore forest dryness and hot, dry, windy weather are the necessary and increasingly common ingredients for wildfire activity for most of the summer. Snowpacks are now melting 1 to 4 weeks earlier than they did 50 years ago, and stream-flows thus also peak earlier."

Westerling *et al.* found that, in the 34 years studied, years with early snowmelt (and hence a longer dry summer period) had five times as many wildfires as years with late snowmelt. High-elevation forests between 1,680 and 2,690 m that previously were protected from wildfire by late snowpacks are becoming increasingly vulnerable. Thus, four critical factors — earlier snowmelt, higher summer temperatures, longer fire season, and expanded vulnerable area of high-elevation forests — are combining to produce the observed increase in wildfire activity."[2]

Forest Policies Invite Logging at the Expense of Community Protection

The forest ecosystems of the western United States have evolved with fire, to the extent that some plant species cannot complete their life cycles without periodic blazes. The cycles of burn and regeneration went on for millennia; there's some evidence that Native Americans may have set fires and conducted controlled burns.

Then in 1910, the Great Burn scorched more than three million acres in northeast Washington, northern Idaho, and western Montana over a two-day period, and killed 83 people. That disaster prompted the Forest Service to begin a policy of fire suppression in the western forests. The Forest Service enlisted Smokey the Bear to convince Americans that they alone could prevent forest fires. For decades, the campaign successfully interrupted the natural cycle of low-intensity fires that cleared undergrowth and kept forests healthy. Years of fire suppression created a perfect setup for the most intense wildfires — accelerated by a warmer climate — to demolish everything in their path.

After the devastating forest fire season in 2002, timber-company representatives argued that forests should be more aggressively managed to reduce combustible material that feeds fires once they break out. Forests were becoming "overgrown," and limits on logging made the countryside even more vulnerable to fire, they claimed. In response to the 2002 fire season, President Bush introduced the Healthy Forests Initiative, a series of changes to management of the National Forests. The policy, which was ostensibly designed to "reduce fuels," in fact offered timber companies more access to the National Forests. By allowing logging of large tracts of wild lands, the President's plan encouraged timber companies to build roads and cut large, fire-resistant trees deep in the back country — far from homes and neighborhoods at risk. Environmentalists saw the plan as designed to push aside political opposition to industrial logging rather than to clear the underbrush that clogs the forest floors. Over their objections, the policy was signed into law in December 2003.

In Montana, on the Bitterroot National Forest, the first project ran into public opposition. The Administration's project contained only super- ficial protection from wildfire but also called for clear-cut logging over four square miles (3,000 acres) of old-growth, native forest. These forests of Douglas Fir are home to elk and bighorn sheep, and world-class trout fisheries. They're beloved not just for their biodiversity but their history: Chief Joseph led his Nez Percé tribe here, and Lewis and Clark once explored these forests. No wonder public opposition to the Administration's plan was strong.

Working with Bitterroot Valley residents, conservation groups created an alternative, a plan that included more residential protection, carrying out fuel reduction on 1,600 acres, and would have provided 45 local jobs and pumped $1 million into the local economy. The citizens' plan would have preserved the old-growth forests, as well. Even though 98 percent of the 13,000 public comments on the project favored the conservationists' alternative, the Administration ignored it and went ahead with their logging plans.

Whether or not these policies endure, it is clear that federal decision- makers — faced with worsening fire seasons and the threat of climate change — missed an opportunity to reform years of mismanagement of the national forests. Meanwhile, back in the woods there's an even more ominous trend afoot, one that comes with much higher cost and greater risk.

Development near Forests Risks Lives, Busts Budgets

Drawn by the beauty of nature and opportunities for recreation, people are moving to the West, and building homes on the edges of National Forests, in the "wild-land urban interface." Many newcomers have relocated from places where wildfires are not common, and know little about the threat.

In these new suburbs on the fringes of national forests, the risk of fires is rising, and so are costs. Protecting new homes from wildfires is straining the Forest Service's budget to the limit. Forest Service managers estimate that safeguarding these privately owned properties will soon cost the agency $1 billion per year. Headwaters Economics, a nonprofit research firm based in Montana, has studied the potential severity of the problem. Their analysis reveals:

Only 14 percent of forested western private land adjacent to public land is currently developed for residential use. Based on current growth trends, there is tremendous potential for future development on the remaining 86 percent [See Figures 2.2 and 2.3.].

One in five homes in the wildland urban interface is a second home or cabin, compared to one in twenty-five homes on other western private lands [See Table Figure 2.4]. Residential lots built near wildlands take up more than six times the space of homes built in other places. On average, 3.2 acres per person are consumed for housing in the wildland urban interface, compared to 0.5 acres on other western private lands.

Given the skyrocketing cost of fighting wildfires in recent years (on average $1.3 billion each year between 2000–2005), this potential development would create an unmanageable financial burden for taxpayers.

If homes were built in 50 percent of the forested areas where private land borders public land, annual firefighting costs could range from $2.3 billion to $4.3 billion per year. By way of comparison, the U.S. Forest Service's annual budget is approximately $4.5 billion.[3]

In Montana, the 2007 wildfire season stretched the resources of the state beyond its ability to pay. In late August, with the wildfires still raging, Governor Brian Schweitzer called lawmakers into a special session of the Montana Legislature to set aside $55 million to fight the wildfires that had already burned more than 400,000 acres. Montana's share of the state firefighting bill was $35 million, $19 million more than all funds set

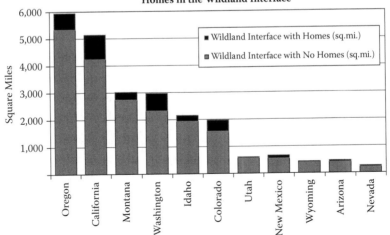

FIGURE 2.2   Homes in the Wildland Interface.

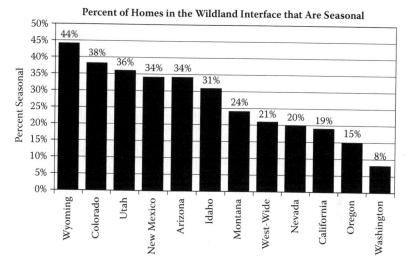

FIGURE 2.3   Percent of homes in the Wildland Interface that are seasonal.

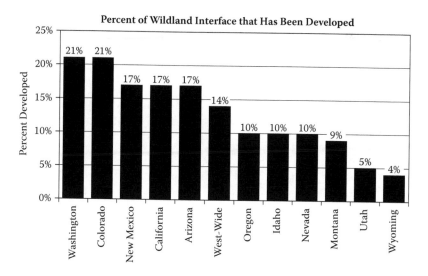

FIGURE 2.4   Percent of Wildlife Interface that has been developed.

aside to pay for disasters for the next two years. Even if the state spent every dollar of the governor's $16 million emergency fund, and exhausted another $8 million from other budgets, $11 million of the state's firefighting bill still remained unpaid — and the fire season wasn't over.

My own experience, serving as regional director of the Red Cross in Montana, mirrored the struggles of state officials. The Red Cross responded to nineteen forest fires from June to September; during this time, the chapter spent its entire disaster budget, leaving no funds to respond to other emergencies, including house and apartment fires or any other disasters. Staff members were exhausted, and the state chapter had to appeal to the national organization for help with funding. The fires raged until snow and rain in the high country finally extinguished the blazes in September.

Climate Change Brews Bigger Storms, More Flooding

Scientists predict that climate change, even at moderate scenarios, will make floods a common phenomenon. Along the coasts, where half of our population lives, storms will be driven by warmer ocean temperatures and hit harder than before. And melting ice caps translate into higher sea levels and more coastal flooding. According to the International Panel on Climate Change, global sea levels have already risen 6.7 inches.[4] Inland areas will not be immune: a warmer climate means less precipitation falling as snow and more heavy rainfall, which runs off the land more rapidly, again causing flooding. In the arid West, lands burned by intense wildfires are especially vulnerable to subsequent flooding when compromised soils cannot absorb rainfall.

Wetlands Are Nature's Sponges during Floods

The U.S. Geological Survey defines wetlands as, "areas that are wet due to a close relationship to a body of water or groundwater, or land areas that are flooded regularly; they support vegetation adapted for life in saturated soil conditions." Swamps, bogs, fens, tidal marshes — we know these various ecosystems collectively as wetlands. Settlers arriving in the New World viewed these places as dangerous, full of mosquitoes and air infected with "miasma" that spread disease. The impulse to conquer wetlands was strong; according to the Environmental Protection Agency, half of the wetlands present before settlement have been destroyed. When the government started tracking the status of wetlands in the 1980s, the country was destroying about half a million acres of wetlands each year — an area that's about the size of Rhode Island.

Now there's an awareness that these natural systems serve as stores for groundwater; they filter sediments, purify water, provide homes for wildlife, produce timber, fish, and shellfish and give us places for recreation, tourism, and cultural values. And it is also obvious that wetlands have a specific function when it comes to climate change and disaster mitigation. Wetlands "hold" and slowly release floodwaters; they stabilize shorelines, and in the event of a storm surge in coastal areas, wetlands break the force of destructive waves, mitigating the deluge. Whenever there's too much water, wetlands act as natural sponges.

The Bad Math of "No Net Loss" of Wetlands

Once the "ugly ducklings" of the natural world, wetlands are now respectable enough to be mentioned during presidential campaigns. During his 1988 run for office, George H. W. Bush promised "no net loss" of our nation's wetlands; not to be outdone, Bill Clinton pledged a "net gain," and following in his father's footsteps, George W. Bush also committed to the "no-net-loss" program. Yet during each administration, although the pace of destruction has slowed somewhat, these natural areas have continued to be degraded, dredged, drained, and polluted.

Then, in March 2006, the Bush Administration claimed that the country had finally achieved the "no net loss" goal, and had even begun gaining wetlands. On the eve of her departure, then Secretary of the Interior Gale Norton released a survey compiled by the U.S. Fish and Wildlife Service that showed a net increase of wetlands of 192,000 acres from 1998 to 2004. Norton announced that after decades of trying to stop the loss of wetlands, the country had finally succeeded. "For the first time since we began to collect data in 1954, wetland gains have outdistanced wetland losses," Norton said. Environmentalists revealed that the Administration had counted so-called open water habitats, or "created" ponds and reservoirs. "They counted building ponds, pits at gravel mines, highway run-off retention ponds and golf course water traps," said Julie Sibbing of the National Wildlife Federation. "Meanwhile, we're still losing natural wetlands."[5]

The principal federal program that protects wetlands is found in the Clean Water Act, written in 1972. Section 404 of the Act prohibits the discharge of dredged or fill material into certain wetlands without a permit from the Corps of Engineers. To help achieve the goal of no net loss, the Corps can require compensatory mitigation such as restoring another, degraded wetland as a condition of a permit when wetlands destruction is unavoidable. Permittees can perform the mitigation themselves, pay a mitigation bank to perform the task, or agree to an in-lieu-fee arrangement.

In 2005, the Government Accounting Office studied the Corps' wetlands mitigation program, visiting seven of the Corps' district offices in diverse locations to review case files. For the 152 cases that required mitigation, researchers found little evidence that monitoring reports were completed or that the Corps conducted compliance inspections. The report found that the Corps performed little oversight on mitigation projects; on thousands of acres, the agency could not even assess whether mitigation was performed at all.[6]

Another sore spot for environmentalists is the Corps of Engineers' nationwide permit program. Set up under the Clean Water Act, this program offers developers and other potential wetland fillers quick okays on projects of up to ten acres. The intent was to cut red tape for small projects, but the nationwide permits, especially a notorious one known only as Number 26, allowed piecemeal destruction of thousands of acres of valuable wetlands every year. When the program came up for renewal in 1996, environmentalists threatened to sue if changes were not made. Other federal agencies waded into the fight, making it clear to the Corps that their permitting process had consequences for others besides developers. FEMA officials pressured the Corps, bringing the connections to light that linked the permit program to accelerated development in hazardous floodplains. EPA officials also weighed in. Eventually, the Corps agreed to phase out this single destructive rubber-stamp permit. Despite repeated court challenges, environmentalists have prevailed, although the overarching nationwide permit program remains, overseeing what one wetlands advocate characterized as "the orderly destruction of our nation's wetlands."

Other federal programs have been written with the promise of conserving wetlands, but their legacy seems as troubled as the nationwide permit program. During the 1985 Farm Bill, Congress enacted "Swampbuster," legislation that stipulated that farmers who drain wetlands would be ineligible for federal farm program benefits. By withholding payments, the program became the primary line of protection against continued drainage of wetlands on agricultural lands. Environmentalists had high hopes for the program when it was first authorized, but my own experience as a lobbyist during the reauthorization in 1990 left me wondering about the program's effectiveness. In 2003, a study by the Government Accounting Office revealed that the Department of Agriculture had failed to enforce the law. The report concluded that, "almost half of Natural

Resource Conservation Service's field offices are not implementing one or more aspects of the conservation provisions of the 1985 act as required."[7]

Development in Floodplains Risks Lives, Busts Budgets

The Flood Control Acts of the 1920s ushered in an era of taming America's great rivers. Building levees, dikes, and dams and paving over wetlands served the shipping industry and farming on river bottoms proved productive — in the short-term. But these artificial restraints also promoted reckless development of the floodplain. Now, there's mounting evidence that our love affair with structural solutions like dams and levees has actually worsened floods. That's because the natural function of the floodplain and its associated wetlands is to carry excess water during times of heavy runoff. Rivers that have been narrowed down with levees often talk back; the river rises even higher to compensate during times of floods. Decades of straitjacketing rivers, taming them with levees, dikes, and dams that have cut them off from their wetlands-studded floodplains, is like an overdue bill that we're paying with interest.

"As the major flood control projects grew, less and less thought was being given to the notion of whether flood control is indeed an enterprise worth undertaking at all," said David Conrad, Senior Water Resources Specialist at the National Wildlife Federation. Conrad is author of *Higher Ground*, a report that documented how the National Flood Insurance Program (NFIP) was failing to manage both the costs and growth of flood-related risk. The report showed egregious examples of "repetitive losses," cases in which homes or other structures were constructed and reconstructed repeatedly, without regard for risk. The report showed that the NFIP offers little incentive to move out of harm's way, in part because insurance rates were below (some of them far below) true actuarial rates. In addition, the program used flood-hazard maps that were inaccurate or outdated and failed to consider changing conditions. The Wildlife Federation's report also blamed local communities and FEMA for failing to enforce even minimum standards of the program or set more rigorous standards of their own.

Conrad and his colleagues worked with the Clinton Administration and congressional reformers to overhaul the program. Leaders at FEMA, eager to re-orient the agency's efforts from bailouts and disaster response to preparedness and mitigation, saw an opportunity to make reforms. In 2004 Congress streamlined the program and funded an effort to address repetitive losses. The stage was set for accountability. Then four powerful hurricanes struck Florida in 2004, followed by Hurricanes Katrina, Rita,

43

and Wilma in 2005, and by November 2007 the NFIP had amassed a debt of nearly $20 billion.

The new program is still largely not implemented. In fact, the number of repetitive loss properties has ballooned from 74,500 when *Higher Ground* was published in 1998 to more than 135,000 properties, and the cost to the NFIP of these properties has more than tripled to over $8.5 billion in payments. "The annual interest to the Treasury is $800 million per year, nearly half of its annual revenues. The program has failed to reduce risk, failed to keep people out of harm's way, and it's in fiscal crisis," said Conrad.[8]

Like the failed fire-suppression policy that left the national forests mismanaged and taxpayer dollars misspent, the nation's attempt at blocking the flow of America's great rivers is a losing battle. Floods are getting worse, bailouts are costing more, and the future only looks worse. Mary Landrieu, U.S. Senator from Louisiana, said, "Our wetlands are Mother Nature's levee system, and if the United States does not immediately begin to get serious about investing in these wetlands, I greatly fear that my state is doomed."

Everything Is Different

In 1929, the Hungarian writer Frigyes Karinthy published a collection of short stories called *Everything is Different*. One story, "Chain-Links," explored the notion that rapid travel and modern communications had brought people closer. Although the physical distances remained the same, tighter human networks had created a "small world" by closing the social distance. To illustrate this theory, Karinthy's characters devised a game.

Karinthy writes: "One of us suggested performing the following experiment to prove that the population of the Earth is closer together now than they have ever been before. We should select any person from the 1.5 billion inhabitants of the Earth. . . . He bet us that, using no more than five individuals, one of whom is a personal acquaintance, he could contact the selected individual using nothing except the network of personal acquaintances."[9]

Karinthy's book fell into obscurity and is now out of print, but his idea was rediscovered in the 1960s by Stanley Milgram, who designed an experiment to transfer a letter between people in distant cities using personal hand-offs between friends. Milgram found that people could deliver the letters through an average of 5.5 personal contacts — uncannily close to the five people in Karinthy's story. Milgram is generally known as the originator of the notion of "six degrees of separation," a classic principle of social networks.

Years after Karinthy and Milgram described social networks, our world is even smaller. The new information technologies that connect people more tightly — the cell phones, instant messaging, videoconferences, and online social networking sites, continue to shrink social distance, but these are also the natural tools of democracy that citizens are using to combat climate change. At the grassroots level, through "chain-links" and peer-to-peer communication, networks are taking on the hard task of working toward positive change. The following are a few examples that I find most inspiring.

Networks Restore Resiliency of Forests and Fireproof Homes

Landowners in the idyllic Sawmill Gulch love their neighborhood, which borders the Rattlesnake National Recreation Area, north of Missoula, Montana, but they live in fear of the "big blow-up." Over the years, the character of Sawmill Gulch has completely changed, transforming from stands of big trees punctuated by open meadows to a crowded and dense woodland. For decades, small blazes that once cleared the forest floor have been suppressed.

An unusual partnership between U.S. Forest Service, environmentalists, and timber companies is creating a safer neighborhood by cutting the chance of severe wildfires sweeping into the valley. Concerned about wildfires in the summer of 2003, the groups designed a project to begin timber cutting, brush removal, and small, controlled fires on 754 acres in Sawmill Gulch. The project creates jobs and builds trust among groups that often find themselves at odds.

The project began with volunteers clearing brush on a 98-acre demonstration project near the Sawmill Gulch trailhead, a popular recreation area. As part of this project, the Society of American Foresters and Sierra Club hosted several hands-on workdays, supervised by Lolo National Forest employees, when volunteers removed brush and small-diameter "ladder fuels." Property owners got involved, and some completed fuel-reduction work on their private land. During winter when the ground is frozen or summer when the ground is dry, a commercial timber company, Tricon timber, will cut trees no larger than 25 inches in diameter. The project is expected to take five years to complete, and will reduce the number of trees by up to 50 percent in the work area. After the project is completed, the area will be monitored for two years. "It's about making our forests and our community safer," said Bob Clark, a Sierra Club spokesman.

"It's not often that the timber industry and environmentalists come together, sit down at the same table and agree on a hands-on project that involves the management of our national forests. But in this case, the safety of Montana residents was at stake. In the years to come, we'll see more of this kind of approach. Many citizens want to help reduce the risk of a wildfire burning homes and endangering lives on the edge of our forests. It's clear that these projects help protect fire-prone communities while also invigorating local economies."[10]

When the Missionary Ridge fire blazed through the woods near the Los Ranchitos subdivision in Durango, Colorado, none of the 33 houses burned, and residents think they know why. Their fire-prevention committee, a network of citizens, state foresters, and a company called Fire Ready, had carried out an innovative subdivision-wide plan.

After state foresters conducted an analysis of the health of the woods, Fire Ready removed the ground litter and low tree branches that act as "ladder fuels" or accelerators, thinned the forests, and helped residents make their homes more fireproof. Landowners acted to spread the word about fire protection and forest health. "Everyone knows everyone in one way or another," said David Welz of Fire Ready. "People just naturally connect with each other and create networks to work together and get it done."[11]

Southwestern Colorado has established an "ambassador" program that gets information to homeowners, teaching them to create "defensible space" around their homes. The ambassador creates a network of partner organizations that in turn spread the word about best hands-on practices. The response from citizens has been overwhelmingly positive.

Networks Restore Resiliency of Wetlands, Flood-Proof Communities

Tulsa, Oklahoma, has gone from being a community troubled by repeated flooding to an innovator in the field of holistic, integrated disaster mitigation. In 1964, after the Corps of Engineers completed the Keystone Dam, Tulsa residents believed they had tamed the Arkansas River. Developers paved the hillsides and built in lowland wetlands, and the region's population grew by 25 percent in the 1960s. When flooding did occur, the response was the typical return and rebuild — until the 1976 Memorial

Day Flood that took three lives and wrought $40 million in damages. This time, Tulsa citizens recognized that they had to work at flood mitigation. Tulsa elected new city commissioners who declared a moratorium on floodplain construction. Even so, another Memorial Day Flood in 1984 dumped 14 inches of rain and flooded nearly 7,000 homes, killing 14 people and causing $180 million in damage. In response, Tulsans decided to get even more serious about alternatives.

Leaders in Tulsa established a storm-water protection program with a stable funding mechanism for the maintenance and management of a storm-water utility. The city set up a watershed-wide floodplain management program. Working with the Federal Emergency Management Agency's Project Impact, the city relocated more than 500 houses and 900 other buildings out of the most dangerous zones. Project Impact established a community outreach program to alert residents about flood hazards and offer mitigation solutions and technical assistance for homeowners and business owners. Local officials enacted strong building codes. To restore the natural resiliency of wetlands along the river, the city preserved more than a quarter of its floodplain as open space.

More than 7,000 residents of Napa, California, were evacuated during the flood of 1986, a disaster that caused $140 million in damages. The event was tragic but it was not the first time that the Napa River had flooded. This time residents knew that harnessing the river with more structural alternatives would not solve the problem. They rejected the Army Corps of Engineers' plan to dredge the river and install levees and instead created a "living river" plan. The plan was facilitated by FEMA's Project Impact, and it brought together a broad coalition of community organizations, the Environmental Protection Agency, state and local officials, and planning professionals. Eventually the project will reconnect the river to its wider, historic flood plain, and restore the river's more natural, meandering course. This project will restore 600 acres of wetlands, rich natural areas that can provide habitat for fish and wildlife while also protecting against flooding. More than 60 buildings are being moved from the floodplain. Old dikes will be breached to restore tidal wetlands habitat, and the county plans on purchasing 300 parcels of land, a continuous natural corridor along the river.

Grassroots Models for Positive Change

In two vastly different situations — fighting wildfires and staunching rising flood waters — local partnerships acted to restore natural systems and regain their protective functions. In these examples, local citizens created solutions by bringing together diverse viewpoints. Sometimes they broke with traditional leaders such as city councils or county commissioners; other times, they became actively involved in community leadership. Their community-based solutions were systems-based; they integrated the wisdom of ecologists, engineers, planners, and home owners. The partnerships sometimes dissolved after their goals were accomplished. In most cases, the efforts were heavy on outreach and communication. By restoring forests and wetlands, partners left local wild places healthier than they had found them, strengthening the stabilizing capacities of the ecosystems. Their efforts prove that, in an era of climate change, everyone is a stakeholder in environmental protection. And their positive experiences may expand the constituency for preservation of wetlands and wild forests beyond the cadre of those who explicitly identify themselves as "environmentalists."

Einstein said, "We cannot solve problems by using the same kind of thinking that created them." Under scenarios of climate change, traditional, linear solutions fail to take into account the instability of ecosystems pushed to the brink. In the new, nonlinear world of climate change, more control may be less efficient and less effective. Accounting approaches, in the shape of carbon "trading" and "offset" schemes may prove no more credible than the Bush Administration's "no net loss" of wetlands numbers. But old approaches die hard. Even as this textbook goes to press, environmentalists work diligently at the federal level to defeat short-sighted projects such as the Yazoo Pumps, which would drain more than 200,000 acres of rich bottomland hardwood wetlands in Mississippi, in the guise of flood control.

Nor can networks at the grassroots level replace the need for smart federal policy. We need sensible national policies that treat our wildlands with respect, that are fiscally responsible, that stop subsidizing disaster. We can only prevent climate change with limits on emissions, but we must respond to changing scenarios with a mix of strategies. For example, to protect densely populated urban areas, structural solutions for hurricane protection may be necessary.

At both the national and local level, citizens must press their governments to use comprehensive tools such as planning, zoning, and subdivision control to keep people out of harm's way. These are difficult subjects, particularly in states like Montana. But during the special session of the Montana legislature after the catastrophic wildfires, lawmakers were at least willing to hear from the experts at Headwaters Economics about the implications of growth near our forests, and that's reason for hope. Ultimately, it may prove easier to make adjustments to land-use and community settlement patterns than to negotiate with a 500-foot wall of fire or a storm surge that's topping a levee.

Almost forty years ago, public outrage over polluted rivers, urban smog, and oil spills inspired the first Earth Day. The painful, visible proof of our disregard for nature drove people to demand action. Today, climate change spans borders of cultures and politics but the natural disasters that accompany its arrival speak clearly in the universal language of a planet in peril.

Earth Day 2020 will dawn on a more crowded, more "connected" planet, but I do not believe that it will be six degrees warmer. That's because humans will have responded with courage and creativity and built innovative networks to safeguard the home planet, using a time-tested rule of grassroots organizing: Start where you are.

REFERENCES

1. Personal conversation with Dr. Anthony Westerling, January 14, 2008.
2. *Science* 313, no. 5789 (August 18, 2006): 927–28.
3. Research by Headwaters Economics, Headwaterseconomics.org; personal conversations with Ray Rasker, December 2007.
4. International Panel on Climate Change, 2007, p. 7.
5. Personal conversations with Julie Sibbing, December 2008.
6. GAO Report, 05-898, September 2005, *The Corps of Engineers Does Not Have an Effective Approach to Ensure Compensatory Mitigation.*
7. GAO report, April 2003, *Agricultural Conservation: The USDA Needs to Better Ensure Protection of Highly Erodible Croplands and Wetlands.*
8. Personal conversation with David Conrad, January 2008.
9. Albert-László Barabási, *Linked* (New York: Penguin: 2003), p. 26. Barabasi offers an excellent overview on the origins of social networks. The story by Karinthy is now out of print.
10. Personal conversations with Bob Clark, November, December 2007; January 2008.
11. Personal conversation with Dave Welz.

3

Federal Mitigation Programs
Collateral Stimulus to Reducing the Impacts of Climate Change in our Communities

INTRODUCTION

To date, the federal government has taken an inconsistent approach to dealing with the issues and impacts of climate change. Although it has provided considerable support to investigating and analyzing the issues surrounding climate change, it has been less aggressive in promoting policies and programs that specifically address the impacts of climate change. Instead, certain federal programs designed to support risk reduction, or mitigation of natural hazards, when implemented have demonstrated collateral and unintended benefits in addressing the impacts of climate change on our communities. These programs clearly illustrate that implementing certain mitigation actions in a community will reduce the impacts of future disasters aggravated by climate change.

Federal policy makers have not promoted the connection of these mitigation programs as a potential remedy to climate change. This oversight needs to be corrected. The fact that climate change adds to the severity and frequency of natural disasters provides additional impetus for taking mitigation actions before the next disaster. And an equally compelling case can be made that the mitigation actions taken now will impede the impact of climate change in the future. Making this two-for-one case to

51

community leaders makes economic sense and provides coverage for community leaders that they are addressing the more immediate and politically salient issue of climate change.

The intent of this chapter is to discuss several current and past community-based federal mitigation programs administered by the Federal Emergency Management Administration (FEMA), now a part of the Department of Homeland Security (DHS) and their collateral influences on reducing the impact of climate change. The programs that will be examined include the National Flood Insurance Program (NFIP), the Property Acquisitions program, and Project Impact: Building a Disaster Resistant Community.

In the case of the NFIP, we will explore the controversy surrounding the costs and benefits to communities of participating in the NFIP. The NFIP has been described, on the one hand, as mitigating the future risks from climate change through good floodplain management, and on the other hand, as offering low-cost insurance, which fosters development in coastal areas that exacerbates the impact of climate change.

THE NATIONAL FLOOD INSURANCE
PROGRAM (NFIP): HELP OR HINDRANCE

Jane Bullock

Jane A. Bullock is a partner in Bullock & Haddow, LLC, disaster management consulting firm and is an adjunct professor at the Institute for Crisis, Disaster, and Risk Management at The George Washington University, Washington, D.C. Ms. Bullock has over 25 years of private and public-sector experience culminating in responsibility as chief of staff for the daily management and operations of the Federal Emergency Management Agency (FEMA), the federal agency responsible for disaster prevention, response, and recovery. Since leaving FEMA, Ms. Bullock has worked with a variety of organizations to design and implement disaster management and homeland security programs including the Corporation for National and Community Service, the Annie E. Casey Foundation, the New York Academy of Medicine, the National Academy of Science Transportation Research Board, DRII International, and county and municipal governments throughout the United States.

The NFIP is considered to be one of the most successful mitigation programs ever created. The NFIP was created by Congress in response to the

FIGURE 3.1 Sulphur, LA, January 26, 2006. Ellie Newby talks with FEMA mitigation specialists Braden Allen and John Ormsby and NFIP specialist Tom McDermott about wind-proofing hardware repairs at this mitigation display at Lowe's Building supplies. FEMA puts these mitigation displays in public places to give people who need to build or rebuild choices in making a better building that will resist damage better. MARVIN NAUMAN/FEMA photo.

damages from multiple, severe hurricanes and inland flooding and the rising costs of disaster assistance after these floods. At that time, flood insurance was not readily available or affordable through the private insurance market. Because many of the people being affected by this flooding were low-income residents, Congress agreed to subsidize the cost of the insurance so the premiums would be affordable. The idea was to reduce the costs to the government of disaster assistance through insurance. The designers of this program, with great insight, thought the government should get something for their subsidy. So in exchange for the low-cost insurance, they required that communities pass an ordinance directing future development away from the floodplain.

The NFIP was designed as a voluntary program and, as such, did not prosper during its early years, even though flooding disasters continued. Then in 1973, after Hurricane Agnes, the legislation was modified significantly. The purchase of federal flood insurance became mandatory on all federally backed loans. In other words, anyone buying a property with a Veterans Administration (VA) or Federal Housing Administration (FHA) loan had to purchase the insurance if the property was in a flood

FIGURE 3.2 Pensacola, FL, December 8, 2004. National Flood Insurance Program (NFIP) representative Carl Watts (right) speaks on-air with WUWF News Director Sandra Everhart about the NFIP program for residents of the Pensacola area, many of whom were affected by Hurricane Ivan. FEMA Photo/Mark Wolfe.

prone area. Citizen pressure to buy the insurance caused communities to pass ordinances and join the NFIP. The NFIP helped the communities by providing them with a variety of flood-hazard maps to define their flood boundaries and set insurance rates.

The 1993 Midwest floods triggered another major reform of the NFIP. This act strengthened the compliance procedures. It told communities that if they didn't join the program, they would be eligible for disaster assistance only one time. Any further request would be denied. As a positive incentive, the act established a Flood Mitigation Assistance (FMA) fund for flood planning, flood-mitigation grants, and additional policy coverage for meeting the tougher compliance requirements such as building elevation.

Over the years, the NFIP has created other incentive programs, such as the Community Rating System. This program rewards those communities that go beyond the minimum floodplain-ordinance requirements with reduced insurance premiums.

It is easy to see the value and collateral benefit that the goals of the NFIP, that is, preserving and restoring floodplains, limiting development in flood-prone areas, and implementing better building standards, would have on reducing the impact of global climate change. While all this is true, some critics wonder if the NFIP is actually exacerbating the impact of climate change, because it promotes people living in flood-prone areas, which means that more and more people are projected to be flooded because of sea-level rise by the 2080s. Wetlands, salt marshes, and mangroves are already being impacted by sea-level rise, and development has accelerated this process. Increased precipitation resulting from climate change will increase flooding in most areas.

The argument can be made that if the NFIP were fulfilling its goals, the 25,000 or so communities that participate in the program would already be mitigating their flood risks and reducing the impact of global climate change by preserving wetlands, restoring the natural functions of the floodplains, and minimizing development in fragile coastal areas and other low-lying areas. In fact, there are clear illustrations of where the NFIP is accomplishing just that in communities across the United States. However, there are also examples of where communities have not enforced their NFIP ordinances, where coastal development occurred without NFIP insurance being made available, and where outdated flood maps allowed for unwise and unsafe development into floodplains. Some of these issues have been recently examined as part of ongoing efforts at evaluating the NFIP.

In a study conducted for FEMA by the American Institutes for Research entitled "The Evaluation of the National Flood Insurance Program Final Report," published in October 2006, evidence was cited as to the beneficial impacts that the NFIP has had in mitigation the flood risk, including:

- An estimated 9,000 square miles of the nation's most flood-prone land are protected from future development because they are delineated as floodways to allow for the unhindered conveyance of flood waters.
- At least 6,000 acres of previously developed floodplain land have been returned to open space through purchasing and removing damage prone buildings.
- Over $1 billion in flood damages are being prevented each year.

However, important to dealing with the rising impacts of climate change, the study found that:

- Most flood-prone areas are still subject to being developed in part because the NFIP has no strong provisions to guide development away from floodplains, even those with extreme flood hazards (coastal areas) or valuable natural resources.

FIGURE 3.3 Bown Brook, NJ, September 18, 1999. Darrell Potter Jr. returns home following the flooding in Bown Brook, NJ. Mr. Potter has flood insurance, home contents insurance, and vehicle insurance. Photo by Andrea Booher/FEMA News Photo. Photo Restrictions: Mandatory Photo Credit No fee for Photo.

- Most natural and beneficial floodplain management functions in the United States are still subject to degradation by development, in part because the NFIP has not emphasized the protection of those functions and has few tools to help restore them once impaired.

Another study prepared by Walter A. Rosenbaum and Gary W. Boulare entitled "The Developmental and Environmental Impact of the National Flood Insurance Program: A Summary Research Report" examined the availability of NFIP insurance and its implications for development that would be impacted by climate change.

Among the findings in this study was that a "significant FEMA concern in understanding floodplain development is to characterize the importance of flood insurance, its availability, and its salience compared to other significant considerations in decisions to build or buy property in areas."

Their findings looked at 18 NFIP communities and the survey responses suggested the following:

- A majority of community developers, floodplain administrators, and home owners considered property characteristics and flood-insurance availability to be among the most important factors in decisions about floodplain property ownership.

FIGURE 3.4 Baton Rouge, LA, March 4, 2006. (Left to right) Karl Smith, Joe Sloan, National Flood Insurance Program Specialists, and Leroy Ingram, Hazard Mitigation Advisor, compose a FEMA team reaching out to the public in a mitigation workshop located at Lowe's Home Improvement Center. These workshops are designed to provide information on how to protect homes against future disasters. Robert Kaufmann/FEMA.

FIGURE 3.5 Guerneville, CA. Having successfully raised their home after they lost nearly everything in the January 1997 floods, Robert and Karen Feldt now also carry NFIP insurance. Photo by Dave Gatley/FEMA.

- Property characteristics and the availability of flood insurance were more important than other factors in decisions about purchasing floodplain property.
- While a majority of individuals living in or near a Special Flood Hazard Area (SFHA) recognized their exposure to flood risk, most of these individuals perceived a much lower risk to their own property.
- A large majority of flood-insurance policyholders in and near an SFHA thought it was relatively important to have flood insurance, but a majority would still purchase, build, or stay in an SFHA without flood insurance.

These responses suggest that the availability of NFIP insurance at current premium rates may be an important consideration for some home owners in their decision to purchase floodplain property. However, the survey responses also imply that an absence of flood insurance does not appear a major deterrent to such property purchases, perhaps because many home owners perceive a relatively low probability of flood damage to their property.

In looking at the impact of the NFIP and the prohibition of sale of NFIP insurance within the Coastal Barrier Resource System (CBRS), which was created by Congress in 1982 to address problems caused by development in coastal barriers such as islands, spits, or mangrove trees that shield the mainland from the full force of wind, wave, and tidal energies, their findings were mixed:

- Estimates of NFIP policies issued on CBRS units have been infrequent and major challenges currently exist in creating such estimates. The most recent available data estimates from 2002 show that no more than 4 percent of all CBRS structures — and probably considerably less — were NFIP insured.
- Fragmentary, sometimes anecdotal evidence suggests that the prohibition of NFIP coverage on CBRS property might inhibit development or reduce the developmental rate when compared to comparable non-CBRS properties.

CBRS development is more likely to be constrained when state and local governments collaborate in the process. Available evidence also suggests that many CBRS units have developed, often quite extensively, despite the absence of NFIP insurance. Market forces appear to be an increasingly potent source of developmental pressure on CBRS units as

FIGURE 3.6 Bay St. Louis, MI, December 11, 2007. Preliminary Digital Flood Insurance Rate Maps are displayed at a flood map open house in Hancock County. The event kicks off community adoption of flood maps to meet requirements of the National Flood Insurance Program. Jennifer Smits/FEMA.

undeveloped coastal barrier land becomes increasingly scarce. Finally, inferences about the NFIP's possible impact on development based upon experience with development on the CBRS lands appear to be tenuous.

There is no doubt that the NFIP does play a role in reducing the impact of climate change but it is also clear that it could do so much more. Both of the reports cited above present recommendations for improving the floodplain management and restoration sections of the program, for strengthening the enforcement of ordinances that inhibit building in hazardous coastal and low-lying areas, all of which are being impacted directly by climate change. It also calls for improved hazard mapping that reflects the changing impacts of climate. There is no indication that the administrators of the NFIP are moving quickly to adopt these recommendations.

However, communities that are currently participating in the NFIP don't need to wait for FEMA/NFIP to adopt these recommendations. They can use existing NFIP requirements for building ordinances and the insurance policy cost-savings incentives of the NFIP Community Rating System, which support restoration of the floodplains to adopt and enforce stronger flood mitigation measures that will help them to adapt now to the impacts of climate change and reduce their future losses from natural disasters.

FIGURE 3.7 Biloxi, MI, August 11, 2006. David Clukie and Peggy Johnson from the National Flood Insurance Program talk to visitors at the Governor's Recovery Expo and tell them of the advantages of having flood insurance. Michelle Miller-Freeck/FEMA.

PROPERTY ACQUISITIONS — THE *PERMANENT* FORM OF HAZARD MITIGATION

Fran McCarthy

Fran McCarthy is currently an analyst in the Government and Finance Division of the Congressional Research Service (CRS), a part of the Library of Congress. He principally works on emergency-management policy issues, particularly those related to disaster recovery. He advises Congressional staff and members and works with his colleagues in the development and writing of analytical reports in response to Congressional requests. Prior to his work at CRS, he spent the majority of his career as a civil servant (1979 to 2006) at the Federal Emergency Management Agency (FEMA). At FEMA he worked in a variety of positions, including manager of the Emergency Food and Shelter Program (Title III of the McKinney-Vento Homeless Assistance Act); liaison to the Senate Appropriations Committee; Deputy Director and Acting Director of the Office of Congressional and Legislative Affairs; Chief of the Declarations Unit; and Policy Advisor to the Director of the Recovery Directorate. In 2003 he received a master's degree from the Industrial College of the Armed Forces (ICAF). He is a 1974 graduate of Kent State University.

"Sometimes, Mayor, you have to bite the pickle. Sometime you've got to really recognize that this house by this crik is a mistake!"[1]

— Former Rapid City, South Dakota Mayor Don Barnett addressing a colleague in Tulsa, Oklahoma following the flooding in Tulsa in 1976.

Property acquisitions are indeed considered by FEMA, and many observers, to be the most permanent form of hazard-mitigation measures that can be taken. They can also be regarded as a one of the best tools that communities have to combat climate change through clearing of the floodplains of unsafe structures, restoring wetlands that have been developed, and moving people and structures out of low-lying areas where flooding is being exacerbated by climate change.

Such an action removes properties and people from vulnerable situations and also has a restorative effect on the wetlands that can benefit a large area. While the most permanent does not necessarily connote the most popular, the acquisitions are a key element of most plans to reduce the risk of flooding in a community and to adapt to climate change. This discussion of the national mitigation program focusing on property acquisition illustrates how it works best with a partnership in which federal funding is combined with local leadership to deliver solutions to our communities. The best way to describe how this program can work in a community is through a series of discussions about how individual communities have accomplished this process and the cost benefits to those communities. While most of these benefits are assigned to reduction in future disaster costs, a clear collateral benefit to reducing climate change can be easily understood and assessed.

From 1993 through 2008 acquisitions have proceeded at a steady pace, occasionally accelerated by major disasters that resulted in increased funding for acquisitions. As of December of 2007, the Federal Emergency Management Agency (FEMA) estimates it had contributed to the acquisition of 32,434 properties in all 50 states (the total also includes three in Guam). That total ranges from one property acquisition in Utah to well over 4,000 buyouts in both Missouri and North Carolina. The federal share of dollars obligated for the acquisitions was more than $1.5 billion.[2] The federal share included funds from the President's Disaster Relief Fund (DRF), which is the account through which Congress appropriates the great majority of federal disaster assistance administered by FEMA. But in the case of buyouts of property, other major sources of funds are the Community Development Block Grant (CDBG) at the Department of

FIGURE 3.8 Tracy, MO, May 10, 2007. Flood water covers the Platt County Fairgrounds. Several areas of the City of Tracy received flood damage during the previous week but it could have been much worse if it were not for the federal/state buyout program. FEMA/Teri Mayer.

Housing and Urban Development (HUD) and the U.S. Army Corps of Engineers (USACE). These federal funds are combined with a 25 percent state and local share to pay the costs of acquisitions.

The great majority of the acquisitions have been accomplished through Section 404 of the Stafford Act (Public Law 93-288, as amended), the Hazard Mitigation Grant Program (HMGP). This program is the successor to smaller efforts during the 1970s that targeted, on a micro-level, property acquisitions for properties located in the floodplain and vulnerable to frequent flooding events. HMGP provided greater authority, resources, and flexibility to attempt varied forms of mitigation to reduce the risk to both lives and property of major disaster events.

Other mitigation programs may result in property acquisitions. Three of the other programs at FEMA include the Flood Mitigation Assistance Program (FMAP), the Pre-Disaster Mitigation Program (PDM), and the Severe Repetitive Flood Loss Program (SRFL). The commonality among these programs is that, unlike HMGP, they are not dependant on a disaster event for their funding. Instead, they are prospective programs with an annual appropriation and a planning process that addresses the long-term

FIGURE 3.9   Arnold, MO, May 28, 2003. A federal buyout property is leased to the Jefferson County Youth Organization; the pee-wee football team calls it home. Photo by Adam DuBrowa/FEMA News Photo.

flooding hazard (and in the case of pre-disaster mitigation, other hazards as well) in individual states and local jurisdictions.

The three other mitigation programs are important for establishing priorities that can then be used when a disaster occurs and the HMGP program is activated. Depending on the size of a disaster, HMGP, which can be funded at as much as 20 percent of the disaster total (depending on the status of the state's hazard mitigation plan), can provide a funding pool that allows large project buyouts to be accomplished. Perhaps it provides some perspective to note that of the more than 32,000 properties acquired, over 80 percent of those buyouts were made with HMGP funding.[3]

HMGP can address all types of disasters, including earthquakes and tornadoes as well as flooding events, depending on the threats that face various states and communities. But it is the flooding events that provide the greatest rationale for property purchases. Dating the modern birth of the HMGP program is not coincidental but directly related to the great Mississippi river floods of 1993. At the time that floodwaters inundated the Midwest, the cost-share for the program was 50 percent federal with 50 percent of the costs to be paid for by state and local governments.

In reaction to the floods, and to encourage a vigorous property acquisition program, Congress passed the "Volkmer Amendment" (Public Law 103-181) to the Stafford Act, which increased the federal share to 75 percent and reduced the state and local share to 25 percent. The amendment

FIGURE 3.10 Cape Girardeau, MO, March 27, 2008. SBA Representative Olivia Humilde studies before and after flood photos of an area that participated in a FEMA buyout program in 1993. Andrea Booher/FEMA.

was named for then Rep. Harold Volkmer of Missouri, who represented a portion of the flooded area. His state has gone on to be the most active participant in the buyouts of any state in the nation, with nearly 4,900 properties acquired at a federal cost of $51 million.[4]

While the logic of removing a property from an officially mapped flood-plain may seem obvious, logic does not always equate to the benefits that the government wishes to see from its investment. However, the HMGP program, as well as other federal acquisition programs, have been managed based on cost-benefit ratios that ensure that such benefits can be demonstrated for projects, if not necessarily each individual property. In fact, the program generally pursues several less quantifiable benefits as pointed out by a federal/local study of the property acquisition program in Birmingham, Alabama. The designated objectives of the program included:

1. Ensure fair compensation to home owners experiencing severe economic hardship in the high-risk flood plain of Village Creek in Birmingham.
2. Improve floodwater discharges by removing structures from the floodway/flood plain.
3. Convert such lands into permanent open space, thereby reducing flood levels and the consequential threats to remaining residents of the floodplain.

FIGURE 3.11 Lewes, DE, May 3, 1999. Town signed an HMGP agreement with FEMA to allocate funding for mitigation projects such as buyouts and elevations. Photo by: Liz Roll/FMEA News Photo.

The analysis of the Village Creek project goes on to show that the buyouts, which cost the city and FEMA $7 million, had succeeded in avoiding direct losses of $3.4 million in less than two years. In the same area, a five-year acquisition project by the U.S. Army Corps of Engineers that bought out more than 900 properties helped to avoid "losses of more than $60 million on an investment of $22 million." The report calculates this as "a greater than 150% return on investment."[5]

Effective property acquisition programs require something often difficult to quantify: political will. The low rate of participation in HMGP projects prior to 1993 is partly attributable to the 50/50 cost-share, which made it less inviting, and affordable, for a state or local government to contemplate. But flood insurance regulations and FEMA's regulations would have pointed to a more vigorous program following flooding events, if the FEMA leadership had been willing to enforce its own regulations as noted earlier in this chapter. Federal leadership often provides the cover and support to achieve the necessary political will. This summary of the controversy over buyouts due to flooding in Tulsa encapsulated one view of the situation:

> Patton [Ann Patton, Flood plain manger for the City of Tulsa] and her colleagues knew, based on their disheartening past experience, that they had three days at most to alter victims' expectations when they filed claims with the Federal Insurance Administration to rebuild their

FIGURE 3.12   Crystal City, MO, May 17, 2002. Water stands in a former residential area that state and local officials included in a floodplain buyout program after the 1993 floods. This is the western edge of this buyout area. Photo by Anita Westervelt/FEMA News Photo.

flooded homes. The federal and local bureaucracies of FEMA and the National Flood Insurance program (NFIP) were geared to help victims rebuild their homes and lives as rapidly as possible — despite federal regulations requiring local authorities to abide by salvageable guidelines laid down by the NFIP and not to rebuild "substantially damaged" buildings that did not comply with mitigation requirements. Regarding this regulation Ann Patton said, "We intended to comply with that requirement, even though FEMA was not enforcing it." In the end, the city purchased 306 homes using federal and local funding.[5]

By adding to the equation the real benefit to climate change from property acquisitions, it gives elected officials another, and in some cases more politically acceptable, rationale for pursuing buyouts.

The buyouts following the Mississippi floods of 1993 are now recognized in retrospect as the birth of the revived property-acquisition program. But up close, at the time, it looked very different. As one expert observed: "The buyout program was not universally welcomed. Many residents in the flooded areas viewed the buyout program as a way to force homeowners out of the floodplain and to leave it for the 'tree huggers.' Informational meetings held to explain what the county [St. Charles

County, Missouri] was proposing often became confrontational. County staff attempting to facilitate such meetings often feared that the attendees were merely one step away from physical violence."[7]

It is these types of considerations that eventually and inevitably force the property-acquisition initiative into the "takings" argument. The concept of "takings" has been defined as follows: "When the government acquires private property and fails to compensate an owner fairly. A taking can occur even without the actual physical seizure of property, such as when government regulation has substantially devalued a property." Given the prominence of this argument, as well as fundamental notions of social justice for all, property acquisitions are very much a collaborative effort carried out on a voluntary basis. The decisions begin at the neighborhood level and advance to the city and the state. No jurisdiction wants to create a "checkerboard" of occupied and vacant lots. A buyout that created such a debilitating mosaic would leave some families at risk while also requiring the same amounts of local government support to provide essential services to the remaining residents of the area.

Ultimately, any property-acquisitions program is a series of individual decisions. And they reflect as well on the community's cohesion and sense of itself. While more thought is given to it today, at the time of the 1993 buyouts in St. Charles County, Missouri, it was noted that:

> "No federal program sought to assure that displaces were safely relocated in suitable housing. The buy-out program was specifically exempted from the Federal Uniform Relocation Act of 1970, which would have required relocation expenses to be paid. At this point, no one knows the fate of the former occupants of the flood-prone properties that were acquired."[8]

The St. Charles, Missouri, experience stands in contrast to the experience in Lenoir County, North Carolina, six years later in 1999. The City of Kinston had been flooded over a three-year period by hurricanes Fran, Dennis, and Floyd. After each flooding experience Kinston, working with the State of North Carolina and FEMA, purchased more and more of the vulnerable properties. In all, more than 1,000 properties had been acquired by 1999. But the city was determined to keep its neighborhoods intact and that was an important and conscious goal of the acquisition process:

> The residents who were relocated were allowed to move as neighborhoods in order to preserve their social networks, school districts, and overall community spirit. Ninety seven percent of the homeowners whose properties have been purchased have elected to stay in Kinston. This effort resulted in minimal disruption to the community's tax base.

In addition, new partnerships have been built. A Green Infrastructure plan has been developed in the floodplain to minimize future damage and improve the quality of life. The Greenway will include historical tourist attractions, educational areas, and recreational facilities.[9]

While for some the decision to agree to a buyout is a painful one, in many instances such a purchase represents the rescuing of a family in an untenable position given the status of their properties. In addition, many other corollary benefits can accrue to such a buyout in a floodplain. As one local county engineer in Georgia remarked:

"People who lived on those properties for a long time were glad to have the opportunity to get out," Higgins said. "They knew they weren't going to get anything out of the house. Many of them had sustained a lot of damage, and they would have to disclose it when they sold. We could have carried out a capital improvement project in that area but the cost of such a project would have been at least twice the cost of buying the homes outright. It's nice to be able to regain the storage capacity of the floodplain again. There's also the benefit of re-establishing the natural vegetation and the filtering abilities of the vegetation."[10]

Property acquisition is the permanent form of mitigation and there is some evidence that there is permanence to be gained from the results as well. Throughout, the travails of St. Charles County, Missouri, have been chronicled: from the Volkmer amendment, through potential fisticuffs, to Missouri leading the nation in purchased properties removed from the floodplain. From all of these travails comes a report from the Missouri Emergency Management Agency:

No one imagined that Missouri would get an opportunity to test its buy-out program's effectiveness so soon when the flood of 1995 struck. The 1995 flood was the third largest flood of record in many places, despite the fact that it was considerably less devastating than its predecessor two years earlier. More importantly, the buy-out program resulting from the 1993 flood had removed 2,000 families from harm's way by the time the 1995 flood struck.[11]

It is estimated that 95 percent of the previously purchased properties would have been inundated by the 1995 flooding event. A comparison of the expenditures for the two floods in St. Charles County underlines the savings between the two floods. In 1993, there were 4,227 applicants for supplemental federal assistance, while in 1995 only 333 applications were received. More dramatically, in 1993 FEMA program expenditures exceeded $26 million, but in 1995 those same programs reported spending of less than $300,000.[12]

Because of the buyout not only were funding, people, and property saved but flood waters now had a place to go, and the impact of sea-level rise will be absorbed by this open space, allowing the community to adapt to climate change.

Looking to the future of property acquisitions, the Severe Repetitive Flood Loss Program is attempting to address a vexing problem: disasters often neglect to occur in areas of repetitive flood loss. This small program, with its listing of properties continually making claims against the NFIP, can help to inform buyout decisions during major disasters and point the way for state and local initiatives.

One disaster that did occur in a repetitive flood loss area was Hurricane Katrina. Such a catastrophic event of historical and national consequence, occurring in one of the most flood-prone states in the nation with a large number of repetitive flood loss properties, would appear to be a rich and promising arena for a grand federal/state/local collaboration for mitigation, and with it property acquisitions. But that has not occurred.

The mitigation mantras of building back safer and better have not been heard at the national level following the largest disaster event in the nation's history. State and local leadership, reeling from their own losses and leery of mitigation plans that could, and likely would, have an impact on large portions of the affected area, have taken only small steps to address the future risk. The state plan to rebuild or buy out homes on a vast scale, called "The Road Home," has been plagued by difficulties in planning and funding on such a vast scale. Large amounts of CDBG funds have been appropriated by Congress, but the state and federal leadership have disagreed on fundamental questions of eligibility that have resulted in funding shortfalls and myriad disagreements. Bold federal leadership is not the only answer, and can be inappropriate for some problems. But in this instance, its absence has been telling, and there has not been an effective partnership established between the federal government and local leaders.

Property acquisitions remain not only a permanent, but also a vital and important part of most mitigation efforts that can have a dramatic impact on our communities dealing with climate change. Its history is, ironically, checkered, but now appears steady. The necessary elements are to create a partnership that provides federal funding and incentives that allow local communities to apply all of the tools available from the federal government with local leadership, consensus building, and political will. Engaging in

property acquisition as a positive initiative to deal with climate change as
well as reduce the impact of future disasters is a powerful and politically
viable argument for each community leader.

PROJECT IMPACT:
BUILDING A DISASTER-RESISTANT COMMUNITY

Brian Cowan

Brian Cowan currently works in the Department of Homeland Security,
and since 2001 he has directed a financial-assistance program for local fire
departments that has awarded nearly 40,000 grants for nearly $4 billion.
In 1976 he began a career in hazards management with an assignment
in the National Flood Insurance Program. In 1979 he entered the newly
formed Federal Emergency Management Agency (FEMA) and worked in
various natural hazards programs until 1994, when he joined the FEMA
Director's Policy and Regional Operations office. In 1997 he became the
FEMA director's mitigation policy adviser. Mr. Cowan graduated from
Georgetown University in 1971.

Project Impact (PI) was an initiative of the Federal Emergency
Management Agency (FEMA) to reduce the impact of disasters, in particular natural disasters. It was generally viewed by those associated with or
involved in the effort — state emergency management agencies, state and
federal governmental entities, local jurisdictions, interest organizations,
businesses, academia, and individuals — as a worthwhile and (to a greater
or lesser degree, depending on the location) successful initiative.

Although Project Impact focused on disaster reduction, it is clear that
the PI process a community engages in to identify its risks provides an
opportunity for a community to acknowledge and address climate change
as an ever-growing risk in all of our communities. The discussion of the
Project Impact imitative will look at how the initiative was implemented
with minimal federal input other than seed funding and will highlight
how important local leadership and the private sector were in making
the program a success. Adapting our communities to climate change will
require a renewed effort at a partnership between the federal and local
government and the private sector.

In establishing the Project Impact initiative, FEMA attempted to
learn from the successes and shortcomings of prior mitigation programs,
some of which are detailed earlier in this chapter. The agency also took a

FIGURE 3.13   VA, 1998. Federal, state and local officials gather together at Supply One, a Project Impact partner, to mark the role Supply One is taking in assisting local residents. FEMA News Photo.

deliberate and iterative approach to development of the initiative, building a framework that would reflect the experiences, needs, and goals of the full range of organizations, interest groups, and disciplines that are involved to some capacity in natural disasters. Last, the agency was being directed at the time by James Lee Witt, who placed his personal attention and focus on Project Impact, raised both the internal and the external importance of the effort, and infused into it a marketing approach that proved to be very successful.

The early 1990s included several significantly large impact disasters. Hurricane Andrew brought an unexpectedly high loss to Southern Florida. The loss was tragically high. Yet the surprising aspect of the event was the economic loss; so high that it threatened stability of the home owners' insurance market there. After an administration change in early 1993, Witt, the new FEMA director, was confronted with a huge flooding event on the Mississippi, one that impacted the entire upper and central Midwest. Then, in January 1994, Southern California experienced an earthquake in the Northridge area of Southern California. Again, the economic loss was astounding. A repeat of the Midwest flooding occurred in 1995, along with several landfall hurricanes.

In all of these events (with the exception of Hurricane Andrew), FEMA carried out, and was perceived to have carried out, an effective

FIGURE 3.14 Washington, DC, February 26, 1998. FEMA Director James Lee Witt and President Clinton discuss Project Impact initiatives at a news conference. FEMA News Photo.

and proper response to the events. This is an important prerequisite for Project Impact because it created an acceptance of FEMA as authoritative when it discussed emergency management issues, a perception that until 1993 FEMA did not truly have; and, from the perspective of the discipline of emergency management, it enabled a focus on mitigation of natural disasters. An emergency-management entity that cannot adequately carry out its core responsibility to save lives and property in the immediate aftermath of an event will never be able to adequately carry out an effort to save lives and property before an event occurs.

Important to the Project Impact initiative was a real increase in the number of severe weather events — including winter storms, tornadoes, and hurricanes — and the increase in the human and economic cost of all disaster events. While such losses are clearly not considered positive, they did begin to instill a desire to do something about the escalating losses. Such awareness is central to any social change effort, such as the mitigation initiative. Climate change and the impacts of climate change on our communities now register this same type of awareness, if not an even greater awareness than natural disaster. Unlike with a disaster, where people still think, "It might not happen here," every day people see examples of climate change, real or imagined, in their weather or in the media attention to loss of habitat for the polar bears. People and communities are beginning to look for a way to deal with the still-amorphous risk of climate change. The Project Impact process provides an avenue for action.

In developing an effective disaster mitigation effort, FEMA and its director were aware of and needed to wrestle with several inter-related questions, most of which had been long-standing, and a few of which had risen to prominence during the aftermath of large disasters that arose in the early 1990s. Among these are:

- Who is responsible for paying for the reduction of disaster losses; what is the reasonable extent to which this responsibility can be shared?
- Who are the appropriate partners for sharing the cost of hazard mitigation?
- What is the proper level to which mitigation should be carried out; more harshly, what is the level of acceptable loss?
- What level of accuracy can be achieved with respect to the prediction of disaster losses?
- What is the proper role of government — federal, state, and local — in the implementation of disaster mitigation; what is the proper role of the private sector and individuals?
- Can the probability of a disaster, especially "catastrophic" disasters, be defined?
- Given the transience of government initiatives, and the amount of time it takes to implement long-term mitigation strategies, how does a government initiative endure for the length of time it would take for a mitigation effort to begin to show results?

In order to obtain answers, or at least guidelines toward answers, for these questions, the FEMA director, James Lee Witt decided that he would hold meetings — town meetings — throughout the country and with every discipline, societal component, and emergency management partner: state and local organizations and political representatives; banking and insurance; architects and engineers; hospitals and medical practitioners; business and industry; academia — universities and higher education; social scientists; meteorologists; seismologists; hydrologists; print and broadcast media; other federal agencies; public interest groups; and the general public. In these town hall meetings the director would begin the discussion by pointing out the losses caused by recent disasters. Each of the attendees could relate his or her recent disaster experiences or perceptions. The discussion

FIGURE 3.15 December 10, 1998. The panel for the "Leveraging a Bake Sale into $2 Million" discusses how a modest grass-roots spark can mobilize a community and generate political support for making a community disaster-resistant. FEMA News Photo.

then would turn to the need to do something about escalating disaster losses, and the strategies available to address them.

It is easy to substitute the words *climate change* for *reduction of disaster losses* in each of these sentences and see how PI can help us deal with climate change.

The town meetings were held over ten months or so of 1996. Creative, useful ideas evolved over the course of them — specific suggestions for implementation of a hazard-mitigation initiative.

Through the town hall meetings, the basic programmatic principles of Project Impact had begun to be defined:

- The initiative needed to be community based.
- It needed to have a partnership element to it that was declared openly for the benefit of all concerned.
- The effort needed to be recognized as a long-term one.
- Government needed to play, at least initially, a leadership role — providing or helping to develop incentives where it could; establishing guidelines for a minimum level of mitigation for communities; and demonstrating its commitment to the importance of the effort.

An ad hoc group of individuals experienced in the conduct of mitigation programs, and in particular the implementation of mitigation, was assembled. At the end of its discussions, the group provided its proposals for a national mitigation plan, along with several recommendations for the FEMA director in his mitigation initiative:

- Implementation of mitigation must be the goal. Without that, the initiative could not be a success. Of course, other elements would be needed, such as public awareness, risk identification, and easily implemented technologies.
- The community is without question the right level at which to carry out mitigation. Disasters are local events. Instilling mitigation into the way a community is operated helps to ensure longevity to the effort. And communities know what their vulnerabilities are better than most.
- Financial incentives are essential.
- A public–private partnership should be formed to address disasters at the community level. This partnership should be publicly adopted. So all elements demonstrate that they take the matter of disaster mitigation seriously.
- The federal sector, represented by FEMA, must be involved, and must be seen to be involved. It should provide financial assistance, focusing its funding on those mitigation targets most likely to decrease the cost of a disaster to the agency's Disaster Relief Fund. The federal sector must also be a source of support for risk analysis and a source for minimum mitigation codes and standards.
- The role of the private sector cannot be overemphasized, or overestimated. Business and industry have several critical roles to play in managing disasters — it typically owns all or large portions of critical infrastructure such as electricity or gas; the lack of involvement by the private sector in disaster response had hampered those actions; and business and industry have a huge investment in the continuity of business — thus its interest in loss reduction is economically motivated. For example, employees whose private lives have been devastated by a disaster impact negatively on the productivity of the business.
- The marketing of the initiative needs to be visible and effective. The initiative should have an element of campaign to it, so that the public, both internal and external to the community, is aware of it.

FIGURE 3.16 Wilmington, NC, October 14, 1999. Debbie Pratt, customer relations representative for Barnes & Noble Bookstore in Wilmington, NC, stands next to a display of children's books on hurricanes. Barnes & Noble is a Project Impact Partner that provides educational material to help make Wilmington a disaster-resistant community. Photo by Dave Saville/FEMA News Photo.

FEMA then designed a pilot effort, which involved a handful of communities with a proportionately large number of different aspects — large versus small communities; a range of natural hazards; geographic diversity; and a likelihood of success. In all of the potential pilots, FEMA wanted to have willingness on the part of the community, and some indication that the community had ideas about getting hazard-mitigation implemented. Working with the FEMA regional offices, a list of approximately

FIGURE 3.17 Wilmington, NC, June 3, 2000. The Project Impact Booth at the "Project Impact Hurricane Preparedness Expo 2000" in Wilmington, NC, on Saturday, June 3. There were more than 120 exhibits at the Expo and more than 8,000 attendees. Photo by Tim Carp/FEMA News Photo.

FIGURE 3.18 December 9, 1998. The panel for the "Bridging the Gap between the Public and Private Sectors" workshop included: Robert Quick, president/CEO Metropolitan Evansville, Evansville, IN; Larry Deetjen, city manager, City of Deerfield Beach, FL; Arrietta Chakos, asst. city manager, City of Berkeley — FEMA News Photo.

50 communities was compiled. In the end, seven were selected: Deerfield Beach, Florida; Seattle, Washington; Oakland, California; Allegheny County, Maryland; Wilmington/New Hanover County, North Carolina; Pascagoula, Mississippi; and the counties of Tucker and Randolph, West Virginia.

Pilot communities were asked to do four things:

1. Create a community partnership that includes all elements of the community; elected officials, private sector, nonprofits, academia, unions, religious organizations, private citizens, and all other organizations.
2. Identify the community's hazards and its vulnerabilities to those risks.
3. Identify and prioritize risk-reduction actions in the community.
4. Communicate what they were doing in their community to make their community disaster resistant.

The important element of each of the pilots was that the community directed its own efforts, utilizing the partners to the extent that both parties could operate together. One community held fund-raisers to obtain mitigation resources; in another, a local banking institution offered low-cost loans; in a third, the community organized volunteers to enter the homes of elderly citizens, and carry out simple hazard-mitigation retrofits. In many communities, federal seed monies were matched and leveraged with support from the private sector over 200 percent. With the experiences and lessons of the pilot, FEMA determined that it would place a Project Impact community into each of the states in the following year, growing the initiative to more than 50 communities. The following year, FEMA looked to reaching five communities in each state. Interest in the initiative grew exponentially. FEMA marketed the initiative, and It began to conduct annual conferences — so that information and lessons could be shared — but it was the communities themselves that began to bring excitement and energy to Project Impact.

FEMA's Project Impact initiative can reasonably be considered successful, since it did result in the implementation of mitigation both directly and indirectly; it grew in size; and was viewed as an attractive program by local jurisdictions. It generated a positive energy for its participants. The community-centric aspect of Project Impact, and its focus on implementation, are generally viewed as the main reasons for the initiative's success. As indicated above, government initiatives are usually at some risk of terminating under the political winds that so often drive them. Yet, in the very few years of its formal existence, the effort more than quintupled in size, from 50-plus to more than 250. Communities wanted

FIGURE 3.19 Wrightsville, NC. Wrightsville Beach Mayor Avery Roberts discusses that new residential building codes established after Hurricane Fran kept new homes out of harm's way during Hurricane Bonnie. No new residential homes suffered damage from the 2- to 3-foot storm surge flooding. FEMA News Photo.

to be a part of Project Impact. It was not far from evolving into the thousands when the country underwent a change in presidential administrations, and with it, a change in political appointments. The director of FEMA changed, and the new director determined that he would not maintain Project Impact.

But the PI communities continued their work. In the post-9/11 environment, many of these communities used the PI partnership and process to look at their risks from potential terrorist events. This simple fact demonstrates how easily the PI process can be adopted and can help people deal with the impacts of climate change in their communities. The formation of a partnership and a community-wide committee to examine the risks and vulnerabilities and then begin to address these risks is just good common sense. Heightened awareness of the increased risks from climate change makes sense. It does not cost a community a lot to undertake the PI process, but, as with other mitigation measures discussed earlier, it does require local leadership and political will. Unlike other programs, the leadership and will can come from the private sector as well as elected officials and their economic interests can drive the process to the benefit of the community since climate change impacts us all.

FIGURE 3.20 WV, June 10, 1998. Partners sign the Memorandum of Agreement, supporting the actions set by the West Virginia community partnership. FEMA News Photo.

CONCLUSION

This chapter has detailed three different options for achieving both the goals of hazard mitigation and climate change adaptation. Within the NFIP model, communities who are currently participating in the NFIP can use existing NFIP requirements for building ordinances and the insurance policy cost-savings incentives of the NFIP Community Rating System, which support restoration of the floodplains to adopt and enforce stronger mitigation measures that will help them to adapt now to the impacts of climate change and reduce their future losses from flooding and hurricane disasters.

Property acquisitions remain not only a permanent, but also a vital and important part of most mitigation efforts that can have a dramatic impact on our communities dealing with climate change. Engaging in a property acquisition program as a positive initiative to deal with climate change as well as reducing future disaster impacts is a powerful and politically viable tool for each community leader.

FIGURE 3.21 Carolina Beach, NC, September 22, 1998. Carolina Beach Mayor, Ray Rothrock, holds a press briefing highlighting his community's dune-stabilization and beach-renourishment program. FEMA News Photo

The Project Impact process can be adopted and help communities to deal with the impacts of climate change in their communities. The formation of a partnership and a community-wide committee to examine the risks and vulnerabilities and then begin to address these risks is just good common sense. Heightened awareness of the increased risks from climate change makes sense. It does not cost a community a lot to undertake the PI process but as with other mitigation measures discussed earlier, it does require local leadership and political will. Unlike other programs, the leadership and will can come from the private sector as well as elected officials and their economic interests can drive the process to the benefit of the community since climate change affects us all.

REFERENCES

1. Robert E. Hinshaw, *Living with Nature's Extremes: The Life of Gilbert Fowler White* (Boulder: Johnson Books, 2006), p. 201.
2. U.S. Department of Homeland Security (DHS)—Federal Emergency Management Agency (FEMA), Mitigation Directorate.

3. Ibid., DHS/FEMA Mitigation Directorate.
4. Ibid., DHS/FEMA Mitigation Directorate.
5. FEMA, U.S. Army Corps of Engineers, City of *Birmingham, Alabama,* Post-Disaster Economic Evaluation of Hazard Mitigation, Losses Avoided in Birmingham Alabama, October 2000.
6. Hinshaw, p. 203.
7. Platt, p. 228.
8. Platt, p. 282.
9. FEMA, Kinston-Lenoir Floodplain Leads to Planning, http://www.fema.gov/mitigationbp/briefPdfReport.do?mitssId=290.
10. FEMA, *Buyouts Cancel Complaints, Cobb County, Georgia,* http://www.fema.gov/mitigation/briefPdfReport.do?mitssId=765.
11. *FEMA Property Acquisition Handbook,* St. Charles County, Missouri, p. viii, http://www.fema.gov/government/grant/resources/acqhandbook.shtm.
12. Ibid.

4

Community-Based Hazard-Mitigation Case Studies

INTRODUCTION

The increased frequency and severity of disasters are two of the principal impacts of global warming. Evidence of this trend can be seen in the spate of tornadoes that have occurred across the United States in late 2007 and early 2008, often striking in communities unaccustomed to tornadoes, such as Atlanta. During the 2007 hurricane season, two Category 5 storms made landfall for the first time since the NOAA started keeping records in 1886. In 2007 a persistent drought in the southeastern United States pitted the states of Georgia and Florida against each other in a battle over water rights.

It is becoming increasingly clear that community leaders around the country must be prepared to deal with the fact that global warming will influence how often and how destructive disasters will impact their communities, their economies, and their environment. The lessons learned from communities that have dealt with chronic disaster risks over long periods should serve as a guide for communities looking to reduce the impact of global warming.

This chapter presents two case studies of communities, Tulsa, Oklahoma, and Berkeley, California, that have taken an aggressive, community-based approach to reducing the impact of their chronic hazard risks. For decades, Tulsa suffered from chronic floods that resulted in loss of life and injury and severe damage to private property, public infrastructure, and the community's economy and environment. Citizen action, community leadership, collaborative partnerships and a shared vision of a safer community drove the effort to make Tulsa flood-resistant.

The city of Berkeley is the home of the University of California Berkeley and sits atop the Hayward earthquake fault. City leaders, the university, and the community-at-large have worked together for years to create, fund and implement a series of community-based hazard mitigation programs designed to reduce the devastation when the next earthquake strikes.

These two case studies were authored by individuals who have been deeply involved in these mitigation efforts, and they offer insights and ideas that community leaders across the country should consider in developing similar efforts to address the negative impacts of global warming in the future.

A TULSA STORY:
LEARNING TO LIVE IN HARMONY WITH NATURE

Ann Patton

Ann Patton is a charter member of the team that built Tulsa's flood-control and hazard-mitigation programs. She was also the founding director of three award-winning local programs: Tulsa Partners, Project Impact, and Citizen Corps, all working through partnerships to create safe, sustainable families and communities. She heads Ann Patton Company LLC, a professional consulting firm. She serves as consultant and/or volunteer with groups such as the Institute for Business & Home Safety, Save the Children, and Tulsa Partners. She has worked with the Department of Homeland Security, the Federal Emergency Management Agency, U.S. Corporation for National and Community Service, the Surgeon General's Office for Medical Reserve Corps, and the U.S. Army Corps of Engineers. She is secretary for the Board of Direction of the national Multihazard Mitigation Council. She has served on the Millennium Center Executive Committee, Disaster-Resistant Business Council, the National Working Group on Citizen Engagement in Health Emergency Planning, the Hazard Mitigation Working Group of the Department of Homeland Security, and the Association of State Floodplain Managers' committee on building public support for local floodplain managers.

Introduction

Stepping gingerly over muck-slicked floors, upturned appliances, soggy sofas, and sodden carpets, survivors in the Meadowbrook neighborhood gathered in Carol Williams's flooded living room. It was June 8, 1974. The latest in a series of Tulsa floods had just flashed down Mingo Creek, directly through their neighborhood — again. People had lost count of how many times the neighborhood flooded since it was built in the 1950s.

The water was down now, but it was dark and dangerous in Carol's living room, a haven for snakes and spiders, floors too slimy to walk, and nobody was sure about the wiring. The air was heavy with the stench of foul water. Carol recalled a woman running through the streets in the night, illuminated by lightning flashes, screaming, "My baby, my baby!" Carl Moose spoke quietly about wrenching his boat from the garage just in time to run his latest flood rescue, now becoming almost routine. Bob Miller said his family spent his daughter's ninth birthday stranded on their rooftop, watching their cat drown, with water lapping to their eaves — again.

Everybody agreed on one thing: We have to do something.

The '74 flood was neither the first nor the last on Mingo Creek. But the group that formed that day began a fight that would, in time, change the way Tulsa does business and would influence, to some degree, the nation's disaster programs, too.

This chapter describes some of what happened in Tulsa and what we learned about ways to build a community that is safe, secure, and sustainable. This chapter includes a bit about the place and characters; about death and disaster, about some of the programs and policies that helped move us forward.

Because this account must be abbreviated, it cannot properly acknowledge the many, many people who dedicated their time and talents to help improve our town. The Tulsa story must begin and end with thanks to these many partners, in our hometown but also from afar, who helped us learn from disaster and turn it into community progress.

Tulsa's Story

Some say a fair amount of human advancement arises in response to tragedy. So it has been in Tulsa.

This section describes how we made our way along, by trial and error, disaster by disaster, to reduce the risks that have plagued our lives since man moved to this locale. It focuses on the years of significant change since the June 8, 1974, flood. Those years could be divided into a series of "eras," and this writing follows that pattern:

1974–1984 — Conflict and confrontation
1984–1990 — Challenge and change
1990–1998 — Integration
1998–2002 — Collaboration and expansion
2002–2008 — Sustainability

A Crossroads Place

Tulsa was born in northeastern Indian Territory, now Oklahoma, after the Trail of Tears, when Lockapoka Creeks camped on a high bank of the Arkansas River. We call their site Council Oak, after a venerable nearby tree.

This is a crossroads place. The town was built on rolling terrain, where the low, timbered Ozark hills meet the plains; at a weather junction where hot, dry air from the west collides with hot, humid southern air and cool northern fronts. We call this convergence zone "Tornado Alley." It is prone to violent storms that can spawn tornadoes and flash floods that barrel down the many creeks that flow into the Arkansas River.

Its early tents, shacks, and dusty streets were peopled by pioneers, wildcatters, and Sooners, folk who made their own rules and lived by a frontier ethic: a man has a right to do what he wants with his land. In 1905, oil was discovered at Tulsa's doorstep, bringing a gush of wealth. The town boomed. Oil barons built a flourishing city with tree-lined boulevards and marble mansions. They established a tradition of fierce civic pride and generous donations to better their community. To this day, all current evidence to the contrary, Tulsans believe they live in the Oil Capital of the World.

The Arkansas River flooded pretty much every year, through the roarin' twenties and into the Depression, with a possible exception of the dust bowl years. Major disasters produced changes. After the 1908 flood, Tulsa changed its form of government to the Galveston-disaster model, the City Commission government. After the 1923 flood, Tulsans produced a landmark drinking-water system and preserved a 2,800-acre open-space park in the Bird Creek bottoms. During World War II, after the 1943 flood, the U.S. Army Corps of Engineers built Arkansas River levees around Tulsa's precious oil refineries. Floods in 1957 and 1959 produced the push that resulted in the Corps' Keystone Dam on the Arkansas River upstream from Tulsa. The Keystone Dam was closed in 1964 — producing community euphoria. Tulsans believed they would never flood again, a fantasy that lasted for many years.

By and large, nature's extremes were viewed as something to endure. The Weather Service logged a tornado touchdown somewhere in Tulsa County, on the average, every year during the 20th century; but Tulsans firmly believed an old Indian legend that no tornado would touch down in the city; something about hills to the west. The place also produced killing summer heat and winter cold, floods, and droughts; trouble was a way of life in Oklahoma.

Meanwhile, Tulsa was growing. Many early settlers had favored the high ground, perhaps because they were in close touch with nature or perhaps influenced by Native Americans who tended to honor natural mores. Now homes and businesses spilled over the highlands and down into the bottoms of tributary creeks with names such as Mingo, Joe, Fred, Dirty Butter, Bird, and Haikey — names that would become infamous, in time, as flood followed flood, over and over again: 1957, 1959, 1963, 1968, 1970, and more.

1974–84 — Conflict and Confrontation
By 1974, when Carol Williams convened that neighborhood meeting in her flooded living room, Tulsans had become numb to flooding. Mike McCool, now Tulsa's emergency manager but then a cop, cannot count the times he ripped off his gun belt and dived into a flood to rescue some hapless citizen. "It was just the way life was in Tulsa," he says.

After the Mother's Day flood in 1970, Tulsa joined the federal flood insurance program and promised to regulate floodplain land use — but the city neglected to adopt maps that would have made the regulations work. Flash floods came in rapid succession in 1971 and 1973, followed by four in 1974, dubbed "the year of the flood." The June 8 storm was the shocker: flash flooding and three tornadoes racked Tulsa, shredding the myth of invulnerability and leaving $18 million in damages. I was a newspaper reporter then, trying to make sense of it all, and I could not imagine a worse disaster.

Carol's group named itself Tulsans for a Better Community and began tireless agitation for flood control. They drew in supporters from across the city, including courageous maverick technical experts, such as fiery activist Ron Flanagan, a visionary planning consultant who dedicated his life to stopping Tulsa floods. They succeeded in creating a remarkable pool of expertise on the subject, luring in leading technical experts not only from Tulsa but also from across the country.

Locally, they perceived their enemy to be the Home Builders Association. Enmity reigned. It was the decade called Tulsa's Great Drainage War, as protestors played a clenched-teeth game with development interests. Generally, two steps forward toward stronger flood management were countered by a step or two backward when the next election favored pro-development interests, who dubbed the activists as "no growth freaks."

Tulsans for a Better Community matured into a skilled advocacy group, in part because members did their homework, tried to speak with facts, and knew when to attack and when to thank.

Their advocacy program evolved into four major points:

- Stop new buildings that will flood or make anybody else flood worse.
- Clear the most dangerous of the flood-prone buildings and turn the land into parks.
- Carefully install remedial works, such as channels and detention ponds to hold and convey water, considering the offsite and future impacts of the works, watershed-wide.
- Involve citizens at every point.

Carol Williams epitomized the intense, diverse, and colorful group. Carol's specialty was using surprise, unorthodox techniques. She would identify a favorite dessert of a mayor or a department head and shamelessly curry favor by bringing it on her lobbying visits. It would not be a long stretch to say that she garnered a $150 million Corps' flood project on Mingo Creek with her fabled raisin pies for the congressman's aide. Carol could size up people quickly, usually by analyzing their shoes, and adjust her technique for the audience. She left one nonproductive meeting in disgust, saying, "What could you expect from an entire room of black wingtips?" When an embattled commissioner questioned why group members, mostly young mothers, brought their children to the endless string of flood meetings, Carol retorted: "We're training them to keep after you when we die."

The Memorial Day flood of 1976 struck in the middle of the night, a three-hour, 10-inch deluge centered over the headwaters of Mingo, Joe, and Haikey creeks. The flood killed three and left some $40 million in damages to 3,000 buildings. Enraged flood victims stormed City Hall, and newly elected commissioners, sympathetic, responded with a wave of actions. They imposed a temporary moratorium on floodplain building, hired the city's first hydrologist, Charles Hardt, assigned planner Stan Williams to develop a set of comprehensive policies, began master drainage planning, and gained public approval for the first flood control bond issue in many years. (Since 1977, Tulsa voters have not turned down a bond issue or sales-tax initiative for flood control, according to Tulsa Budget Director Pat Connelly.)

Within a couple of years, regulation was softened after a pro-development commission came into office, but the main body of the new program held. Although the battles were far from over, in large part Tulsa appeared to have stopped creating new problems. Over at least the next three decades, Tulsans could say proudly that there was no record of flooding in any new building that was constructed in accord with the 1977 regulations.

1984–1990 — Challenge and Change

The 1984 election was another upset. Three of the five city commissioners were sympathetic to flood victims. In fact, the new mayor, Terry Young, had campaigned on a pledge to work on flood issues; and the new Street Commissioner (directly responsible for flood programs), J. D. Metcalfe, was a patrician industrialist who was a member of Tulsans for a Better Community. (I came into City Hall as J. D.'s aide, by the way.)

They had been in office 19 days when the worst flood hit on Memorial Day 1984, killing 14 and leaving $183 million in damages to 7,000 homes and businesses. We huddled in the Emergency Operations Center, shell-shocked by reports of Tulsans drowning on lands that had flooded over and over before. Young and Metcalfe vowed right then that things would never be the same — whatever the political cost.

Within hours, we had mobilized a flood-hazard mitigation team. We proceeded with a great sense of urgency. We had learned over the years, disaster by disaster, what we needed to do to seize this moment and execute bold plans. Within days, we had assessed the damage, identified the areas of highest hazard, slapped on a rebuilding moratorium, and identified repeated flooded properties that were candidates for acquisition. One goal was to stop the flooding by clearing the most vulnerable buildings and moving their owners to dry sites. Within 15 days, when FEMA came to town, we were able to meet them at the door with our plans in hand and ask for help to fund them.

It was a fight. FEMA didn't want to fund a buyout, then considered a radical, harebrained scheme. Political opponents charged that the buyout was a "bailout" of people who should have known better than to live in a floodplain. We countered that many of the buildings we identified for buyouts were no longer viable; some had flooded to the ceiling five times in six years, and most would continue to be flood-prone even when all planned structural projects were completed someday in the distant future.

Ultimately, we were able to gain approval to purchase 300 single-family homes and 228 mobile home pads. Mayor Young won over FEMA, and the $17.6 budget included $1.8 in federal and $11.5 million in local funds, plus the proceeds of insurance claims for homes we purchased.

Within a year, we had established a Stormwater Management Department to centralize all flood functions, headed by planner/attorney Stan Williams, which was creating a unified local program to manage flood issues. Within two years, we had instituted a storm-water utility fee, a $2 monthly charge on everybody's water bill, for stable funding of maintenance, management, and planning. We conducted aggressive maintenance and public education programs. We held hundreds of public meetings to get citizens involved in master drainage plans for the entire city. Spurred by planners Sandra Downie and Ron Flanagan, we began including recreation facilities, including trails and soccer fields, in flood-control channels and detention ponds, bringing in a new and positive constituency for storm-water management.

Mayor Terry Young lost the spring 1986 election, but Commissioner J. D. Metcalfe was reelected, and the program continued to evolve.

Another flood hit in October 1986, this time on the Arkansas River. The remnants of a hurricane dropped a 24-inch rain upstream of Keystone Dam, forcing the Corps to release upwards of 305,000 cubic feet per second downstream. It was a challenging time. Every major stream in northeast Oklahoma was at flood, including the Arkansas at Tulsa — despite Tulsans' fond belief that the Arkansas would never flood again

At Tulsa, a private levee broke, flooding 64 buildings. Within days, Tulsa dispatched its hazard-mitigation team and cleared 13 destroyed dwellings, helping their owners move to dry sites.

Overall, the management team worked well, minimizing damages and dangers as much as possible. The new system had passed its first big test.

1990–98 — Integration

In the 1990s, Tulsans began to pull together, united in the vision of a flood-free city. Strong leaders successfully campaigned to change Tulsa's city government from the commission to mayor-council form. The change in 1990 meant that leaders such as Commissioner J. D. Metcalfe, who had

FIGURE 4.1 Sandbaggers fight Tulsa's 1986 flood. *Tulsa Tribune* photo.

FIGURE 4.2 Some charter members of the team that developed Tulsa's flood program. Tulsa Partners photo.

championed the change, left City Hall. Action shifted into a new Public Works Department, headed by hydrologist Charles Hardt; and storm-water management slowly became institutionalized into city operations under Hardt's strong leadership. One of Hardt's skills is building bridges among warring groups, and he helped bring adversaries together to jointly build a safer city. Some former adversaries became strong advocates for floodplain management and, eventually, it became a generally accepted element of the city's services.

In 1987, researcher Claire Rubin had reported that Tulsa County had the most (to that time) federally declared flood disasters of any other community — nine in 15 years. Then, in 1992, FEMA ranked Tulsa's flood program tops in the nation in its new Community Rating System program. Tulsans generally understood that this community, which had one of the worst flooding problems in the nation, was becoming a national model, and they were proud.

Interest in floodplain management peaked again in 1993 when the Mississippi River flooded. With FEMA's new interest in mitigation, floodplain clearance became a respected tool. Tulsa stepped up its ongoing floodplain clearance program. Capital packages routinely included modest funding for acquisition, which the city used as local match with FEMA funding for a continuing pre-disaster floodplain clearance program. By the end of the decade, Tulsa had cleared more than 1,000 of its most dangerous buildings from its floodplains, using the open lands for parks, trails, open space storage, and flood control works.

For the first time since at least statehood, the 1990s decade passed with no significant flooding in Tulsa.

1998–2002 — Collaboration and Expansion

In 1997, FEMA director James Lee Witt launched a new initiative named Project Impact, intended to empower local communities to reduce disaster losses. The idea was to scatter some FEMA money around the country, with few strings, and let locals come up with innovative ways to work out hazard-mitigation techniques, to create "disaster-resistant communities." The ultimate goal, Witt said, was to develop public-private partnerships to change the culture, to establish new cultures that value preparedness and mitigation. In late 1998 Tulsa was fortunate to receive a Project Impact grant for $500,000. I became director of the Project Impact program, named Tulsa Partners.

The Project Impact grant extended over three years and allowed us to expand our hazard-mitigation work beyond flooding into other hazards. We focused on windstorms and tornadoes, lightning, extreme heat and drought, winter storms, hazardous materials, and terrorism after the 9/11 attack.

It also taught us the magic of working through public-private partnerships. Most first-responder organizations and major business leaders became enthusiastic participants in Tulsa Partners. In short order, we had a cadre of dedicated partners working on a very wide range of public

FIGURE 4.3 Tulsa Partners management team, 2000. Photo by Ann Patton.

education and demonstration projects. They ranged from our "SafeRoom" initiative to "McReady." (See Sidebar 1.)

These partners were and are amazing — able, selfless, altruistic, interested in working together without personal gain, united by a common goal to build a disaster-resistant community. As James Lee Witt once said, there is something about the Project Impact process that reaches down into your community and brings out the best in the best of your citizens; he was right. As they worked together, they moved Tulsa into a new era of cooperation. (Wonder of wonders, Project Impact even brought us together with the Home Builders Association of Greater Tulsa, who became the best of partners for the SafeRoom initiative and the green-building initiative called the Millennium Center.)

It is really true that, when it comes to disasters, we have outgrown most of the turf building and petty competitiveness in favor of collaboration and partnership.

Another important advance was long-needed pre-disaster multi-hazard mitigation planning, which got under way in earnest around the turn of the century. Shepherded by planner Ron Flanagan, Tulsa's plan was one of the first approved in the nation and laid out a road map for long-term work toward becoming a disaster-resistant community.

When the Project Impact grant expired after three years, the City of Tulsa and various other sponsors continued to fund the program for several

FIGURE 4.4 Artist's rendering of the proposed Millennium Center at Tulsa Zoo, a demonstration project to show how to live safely in Tornado Alley in harmony with Mother Nature. Source: Tulsa Partners

years. In 2000 we had established a 501-C-3, Tulsa Partners Inc., now ably directed by Tim Lovell, which serves as a useful vehicle for mobilizing public and private donations and creating innovative programs.

2002–2008 — Sustainability

Recent Tulsa Partners projects include continuity planning for nonprofits and small businesses, in conjunction with the Institute for Business & Home Safety; disaster safety for children and care providers, in partnership with Save the Children; and public education and planning for preparedness and mitigation.

At the turn of the new millennium, we expanded our scope again. We had long contended that hazard and environmental issues are two sides of the same coin. For examples, disasters generate tremendous waste and losses; and environmental problems are, in essence, slow disasters. If a house blows away or washes away, it is not sustainable. The most recent expansion of our program includes a shift toward sustainability. Our upgraded goal is to build a disaster-resistant, *sustainable* community.

This goal is at the heart of one of Tulsa Partners' current projects. Still in the planning and fund-raising phase, it is named the Millennium Center. A dedicated group is working to build this demonstration house at the Tulsa Zoo to provide fun, family-friendly, hands-on education on how to live safely and in harmony with Mother Nature.

Sidebar I
Example Projects

Here are some examples of Tulsa-based initiatives that may offer lessons for people working to build disaster-resistant, sustainable communities and live in greater harmony with nature. Many of these projects were born in Tulsa Project Impact or share a similar philosophy.

Tulsa Partners

Tulsa Partners Inc. is a 501-C-3 nonprofit program that continues the work begun by FEMA's Project Impact: creating partnerships to build disaster-resistant, sustainable communities. It serves as a catalyst for collaboration in a broad range of programs, generally related to grass-roots disaster management and sustainability. Public-private partners collaborate to accomplish their mission: to advance community goals, enhance quality of life, and create a more livable, safe, and sustainable community, in harmony with each other and nature.

FIGURE 4.5 Tulsa Partners is an open, inclusive group working for a safe and sustainable community. Tulsa Partners photo.

continued

FIGURE 4.6 Volunteers paint murals in shopping centers, hold pancake breakfasts, sponsor special displays and events, and conduct other education and demonstration projects. Photo by Ann Patton.

Over the past decade, Tulsa Partners has fielded more than 300 partners and hundreds of volunteers. This program has received some dozen awards, including several national awards, as well as grants from national and local organizations.

Funded by grants and donations, Tulsa Partners operates through a governing board, advisory committee, and numerous project-specific committees. It specializes in incubating innovative projects, deriving lessons learned, then institutionalizing those projects with other groups and proceeding to explore new ideas. Some of the best programs have been started by partners within their own organizations, sometimes independently and other times in concert with Tulsa Partners. Several of these programs are described in the following paragraphs.

See also www.TulsaPartners.org.

StormReady

As members of Tulsa Project Impact, the National Weather Service Tulsa staff in 1999 created a new program named StormReady. StormReady established preparedness criteria communities should meet to help them survive weather emergencies. For examples, a town would need to establish a 24-hour warning system and emergency operations center, develop a formal hazardous weather plan with trained spotters, and provide public readiness education. When the

FIGURE 4.7 Tulsa Partners is an open, inclusive, and diverse group. Tulsa Partners photo.

criteria are met, the Weather Service will declare the community is StormReady.

The Weather Service quickly took the program nationwide. As of January 2008, 50 communities have been designated StormReady sites.

See also http://www.stormready.noaa.gov/.

SafeRooms

Tulsa lies in the heart of Tornado Alley, but houses generally have been built without basements or other shelters. In 1998, when Texas Tech University developed new technology for building tornado SafeRooms, Tulsa Partners seized the opportunity to popularize them.

SafeRooms are specially anchored and armored closets or similar small enclosures. They can be built in new or existing buildings, inside or outside, above- or below ground, to provide safe shelter in even the most dangerous tornadoes.

With $50,000 from FEMA, Tulsa Partners formed a partnership with the Home Builders Association of Greater Tulsa to create some high-profile demonstration SafeRooms, coupled with an aggressive public-education program. When the disastrous May 1999 tornado hit Oklahoma, President Clinton kicked off a FEMA-supported SafeRoom initiative, which was later replicated in some other states, too.

continued

FIGURE 4.8 Dr. Ernst Kiesling, Texas Tech inventor of the SafeRoom, checks a surviving SafeRoom after the Moore, OK, tornado, 2003. Photo by Ann Patton.

Within a few years, tens of thousands of SafeRooms were built across Oklahoma. They were used successfully in subsequent tornadoes, such as the 2003 tornado in Moore, OK.

See also *Safe Rooms Save Lives,* http://www.fema.gov/library/viewRecord.do?id=2488.

McReady

In 2003, Tulsa Partners teamed up with McDonald's to provide a month-long family-preparedness blitz in McDonald's restaurants. The program, dubbed McReady, became institutionalized statewide under auspices of the Oklahoma Emergency Management Department.

Spring is the worst season for Oklahoma severe weather. The McReady program links emergency managers across Oklahoma with partners such as the National Weather Service and the Oklahoma Gas and Electric Company. Every April, they set up educational kiosks in McDonald's restaurants, inexpensive grids stocked with family-preparedness guides and other storm-safety materials printed by partners. McDonald's stores print survival tips on tray liners and bags. The low-cost program lasts a month, offering information to the thousands of customers who frequent the state's 170 McDonald's stores each day.

See also www.McReady.org.

FIGURE 4.9 Family preparedness is a central focus of Tulsa Partners. Tulsa World photo. Used with permission.

Tulsa Human Response Coalition

Tulsa's nonprofit agencies have banded together to plan for and manage social services during emergencies. The Tulsa Human Response Coalition includes 50 agencies and first-responder groups. THRC goals are to work together to foster collaboration and communication, share resources, and reduce duplication of effort during emergencies. The

continued

FIGURE 4.10 McReady partners kick off their annual blitz of preparedness information, 2004. Photo by Bob Patton.

group facilitates the human-services aspects of planning, preparedness, mitigation, response, and recovery to ensure effective service delivery. Example projects include emergency mental-health services and life-saving intervention during extreme heat and winter storms or other crises.

The following are among the noteworthy advances in the group:

1. A plan for managing spontaneous volunteers during a disaster.
2. A communications and call center connecting callers with social services and other resources through the 211 helpline.
3. A backlash-mitigation plan, an innovative plan to help the community handle a crisis that could result in retaliation against a specific group (such as Muslims after the 9/11 attack).

See also www.CSCTulsa.org.

First Responders

Tulsa first-responder organizations have a broad spectrum of programs, some inspired or encouraged by Tulsa Partners. Collaboration is the norm and the key to success. The Tulsa Area Emergency Management Agency serves as a central coordinator and supporter for many of these activities. First responders meet regularly in several venues, including a standing Homeland Security Task Force, to share information and ideas.

Activities include Community Emergency Response Teams, a police Disaster Response Team, and a Volunteers in Police Service cadre. Tulsa has a stellar constellation of partners and programs related to emergency medicine and public health, generally arising from the Emergency Medical Services Authority (ambulance services), Metropolitan Medical Response System, Medical Reserve Corps, Tulsa Health Department, and university medical programs. One current focus is planning for public-health emergencies and pandemic flu.

See also www.OKMRC.org.

Disaster-Resistant Business Council

Tulsa is strongly committed to encouraging continuity of operations planning for businesses, nonprofits, and government agencies.

In 2007 Tulsa Partners volunteers formed the Disaster-Resistant Business Council to help spearhead continuity planning. The DRBC is chaired by State Farm Insurance executive Dave Hall. It is a national pilot for the Institute for Business & Home Safety's Open for Business® program. Members include the Tulsa Metro Chamber of Commerce, the Association of Contingency Planners, and the Oklahoma Insurance Department. The DRBC supports Open for Business® planning through workshops, public events and education, direct training, and collaboration with other programs. Examples of recent events include disaster-planning workshops for long-term care, hospitals, businesses, and child-care providers.

See also www.IBHS.org.

Save the Children/Tulsa Partners Initiative

In 2007 Tulsa Partners joined with the international Save the Children group to establish a demonstration project. The general goal is to develop and document ways local coalitions can improve child safety in disasters.

With a wide variety of partners, the group is working to provide preparedness training for children and care providers; to provide continuity of care through Open for Business® planning; and to mobilize

continued

101

FIGURE 4.11   Allison McKee is proud of her disaster safety kit provided by Save the Children. Photo by Elaine Perkins.

neighborhoods and the general community in support of children and their child-care centers.

They also developed a model children's annex to the community's Emergency Operations Plan. Tulsa has formally designated child-care centers as critical facilities — that is, safe and secure child care must continue in place even in times of emergency if the community is to function well and recover.

The *Save the Children / Tulsa Partners initiative* will be documented in a guidebook to help other communities learn from Tulsa's demonstration program.

See also www.SavetheChildren.org.

Planning

Tulsa's commitment to hazards planning is perhaps best exemplified by Ron Flanagan, a planning consultant and activist whose dedication to Tulsa extends over more than 35 years. Notable plans include master drainage plans for all watersheds. Specialized plans guide floodplain management, prioritized capital and acquisition projects, protection of critical facilities, and hazard mitigation. Flanagan served as consultant and catalyst for many of those plans, including the City of Tulsa's hazard-mitigation plan; this prototype plan was adopted on November 25, 2002, one of the first in the nation. Mitigation plan

FIGURE 4.12 Tulsa's Centennial Park storm-water detention pond in downtown Tulsa. Photo by Ron Flanagan.

updates include man-caused hazards. A plan for hazard mitigation in historic buildings is under way.

Flanagan and others helped Tulsa expand the concept of disaster management to include broader issues and constituencies. For example, maintenance easements along drainage channels became the backbone of a community recreation trails system, now extending over more than 50 miles of trails with plans for more. Storm-water detention basins are used for open space and sports fields. Tulsa hazards planning trends toward integration into the larger community fabric.

See also www.rdflanagan.com/Tulsa/Tulsa_NHM_book.pdf.

Environmental Protection

Tulsa is greening up in recent years, with a growing commitment to the environment.

The Metropolitan Environmental Trust champions environmental causes in this area. The M.e.t. provides recycling education and services for Tulsa and its suburban communities. The M.e.t. has recycled 100 million pounds of newsprint in 15 years, for example. Funded by local governments, grants, and private donations, the M.e.t. is considered an authority

continued

103

and program catalyst on many environmental issues. The M.e.t's secrets of success include working to build relationships, finding ways so everybody can win, and fostering collaboration, not competition.

Other programs focus on air and water quality. The poultry industry in northeastern Oklahoma and Arkansas has threatened the quality of Tulsa's drinking water; Tulsa has launched a vigorous team effort to protect its excellent drinking water, long a source of civic pride.

Another noteworthy environmental program is run by the private nonprofit Up with Trees, whose volunteers have planted more than 16,000 trees at 400 sites along Tulsa's streets and expressways since the program began in 1976. In 2007 Up with Trees joined with the city to plant trees along Tulsa flood-control channels and in detention basins, too.

See also www.MetRecycle.com and www.UpwithTrees.org.

The Millennium House and Millennium Center

Inspired by Project Impact, in the year 2000 Tulsa Partner Don McCarthy had a dream: to build a demonstration house that would show how to live safely, in harmony with the environment, at a modest price. He called it the Millennium House. Virtually without funding, except for $15,000 contributed by Tulsa Project Impact, McCarthy and volunteers nonetheless got his house built by 2004. It was open to the public for a year before it was turned over to a low-income family who can enjoy utility costs of little more than $100 a year.

McCarthy's Millennium House inspired an even larger idea. Tulsa Partners pledged to build a permanent demonstration house, to show how to live safely in Tornado Alley while also protecting the environment. They named it the Millennium Center. After two years of team building, planning, and integrating hazard and environmental techniques, the planning group is currently raising funds to build and maintain the Millennium Center and its fun, family-friendly exhibits.

See also www.mctulsa.org.

A Disaster-Resilient Community

Tulsa's skill in managing disaster was tested most recently on December 9, 2007, when an ice storm destroyed tens of thousands of trees and threw 75 percent of Tulsans into darkness. The power outage was the largest in Oklahoma history, with more than 600,000 customer accounts without electricity for upwards of a week or more.

FIGURE 4.13 December 9, 2007, ice storm left 75 percent of Tulsans without power. City of Tulsa photo.

In the Emergency Operations Center, Tulsa Mayor Kathy Taylor and her team developed a three-part recovery program:

1. Removing and disposing of more than 2 million cubic yards of debris.
2. Mobilizing volunteers, including church members and 96 electricians, to help with home repairs, restoration of electricity, and debris clearance.
3. Restoring the city's shattered urban tree canopy, in a public-private wave of tree plantings by the city and Up with Trees.

Meanwhile, across the city, neighbors helped neighbors. The Tulsa Community Foundation launched a campaign to raise funds for emergency human needs. The Tulsa Human Response Coalition established a one-stop center, operated by the Tulsa Urban League, to help low-income people with critical needs. And the electric company and community planners turned their thoughts to long-term mitigation measures, including burying power lines.

See also www.cityoftulsa.org/Storm.asp.

These disaster-resilient programs evolved within days and weeks, born from a community habit — a culture, if you will — of collaborative hazard management, developed over many years in Tulsa's search for ways to live safely in Tornado Alley in better harmony with nature.

Sidebar 2
Tulsa Chronology

1900–2000. National Weather Service records show tornado touchdowns somewhere in Tulsa County, on the average, nearly every year during the 20th century.

1908, 1923, 1943, 1957, 1959. Major floods on the Arkansas River at Tulsa. Records show the Arkansas flooded more or less nearly every year from statehood (1907) until 1964, when Keystone Dam was closed upstream from Tulsa by the U.S. Army Corps of Engineers.

1949, 1957, 1959, 1961, 1963, 1968. These years mark some of the early recorded floods on Mingo Creek and other tributary streams in Tulsa.

May 1970, Mother's Day. Floods on Mingo and Joe creeks cause $163,000 damages and prompt the city to join the flood-insurance program and adopt its first floodplain ordinance. But the city failed to adopt maps that would effect floodplain regulation.

U.S. Army Corps of Engineers, Tulsa District, issues Haikey Creek flood report, the first in a series on Tulsa-area problem creeks.

June 8, 1974. Violent storms cause flooding on Mingo, Joe, Fry, and Haikey creeks. At least three tornadoes also ravage the city. Toll: more than $18 million in damages with more than 120 injured.

October 1975. Tulsa hires its first hydrologist, Charles Hardt, and begins developing comprehensive storm-water management policies.

December 5, 1975. F-3 tornado in Northeast Tulsa injures 38.

March 16, 1976. Tulsa storm sewer and runoff criteria adopted.

May 30, 1976. Memorial Day flood kills three and leaves $40 million in damages.

September 17, 1976. Tulsa imposes a building moratorium in floodplains with critical flooding problems until new maps and regulations can be devised. Moratorium lasted two years

1977. Voters approve a bond issue for emergency flood-control projects, the first in many years.

1977–79. Changes in maps and regulations are adopted, including requirements for storm-water detention and specific permits for

106

floodplain or earth changes. Tulsa begins developing master drainage plans to coordinate changes within entire watersheds.

April 19, 1981. Easter tornado in East Tulsa leaves between $75 and $100 million damage, mostly in an industrial area.

May 27, 1984. Memorial Day flood kills 14, injures 288, and causes $183 million in damages. Tulsa imposes a rebuilding moratorium in repeatedly damaged floodplains until an aggressive program buys and clears 300 homes and 228 mobile home pads. Regulations and standards are strengthened and extended to entire watersheds. Tulsa creates a new Stormwater Management Department to centralize and focus flood-control activities.

September 27, 1985. Tulsa establishes a storm-water utility charge on water bills, $2 a month per household. Proceeds are used for maintenance, management, and planning.

October 1986. Arkansas River flood causes $3 million in damages in Tulsa, $67 million in the region. Tulsa buys and clears 13 destroyed houses after that flood.

March 1987. Researcher Claire Rubin finds that Tulsa County leads the nation (to that time) in federally declared flood disasters, with nine in 15 years.

1990. After a change in the form of government to mayor/council, Tulsa integrates storm-water management into a new Public Works Department. Floodplain acquisition programs continue throughout the decade, with Tulsa purchasing a few properties every year, often with FEMA assistance. By mid-1990s, Tulsa has cleared more than 1,000 buildings from its floodplains.

1992. FEMA gives Tulsa's flood program its best rating in the new Community Rating System, giving Tulsa citizens the lowest flood insurance rates in the United States. Tulsa continued to lead the nation in CRS ratings for more than a decade.

April 24, 1993. East Tulsa/Catoosa F-4 tornado kills seven and injures 100.

April 19, 1995. The Oklahoma City bombing destroys the Alfred P. Murrah Federal Building, killing 168, injuring more than 800, and convincing Tulsans that we need to plan for human-caused hazards as well as natural ones.

continued

1998. FEMA selects Tulsa for a $500,000 grant to start a new Project Impact program — to create public-private community partnerships for multihazard mitigation. The goal was to create a disaster-resistant, sustainable community. Project Impact expanded Tulsa's floodplain vision to include other hazards, including tornadoes. A broad spectrum of educational and demonstration projects addresses reducing risk from tornadoes, extreme temperatures, lightning, floods, and other hazards. Demonstration projects include tornado SafeRooms, historic preservation, social services, fire and law enforcement, Citizen Corps, CERT, and terrorism protection. The grant extends over three years, during which time the group creates a 501-C-3 nonprofit named Tulsa Partners Inc. to continue multihazard mitigation programs.

May 3, 1999. West Tulsa tornado. Tulsa Partners expands its focus again, this time working to marry hazard mitigation with environmental protection. They are working to create a Millennium Center at the Tulsa Zoo, with fun, family-friendly exhibits showing how to live safely in Tornado Alley in harmony with Nature.

2004–2008. Tulsa celebrates anniversaries of its 1984 and 1986 floods. At this writing, in 2008, Tulsa can mark more than two decades without a major flood in the city — a dramatic change from the years when Tulsa experienced a flood on the average of every other year or more often. An ice storm on December 9, 2007, causes the state's largest power outage. Tulsa launches an aggressive program to restore power, curb losses, and restore tens of thousands of downed trees. These points of progress are tribute to the community's dedication to reducing risk and creating a disaster-resistant, sustainable community.

Lessons Learned

Here are a few of the lessons we've learned — almost all of them learned the hard way — on Tulsa's journey toward becoming a disaster-resistant, sustainable community:

- Start with a small hub of your very best people — kind, committed, selfless, and statesmanlike. Develop a shared vision. Then build a holistic, inclusive partnership around that heart.

FIGURE 4.14 Flood projects can be beautiful, as shown in the Centennial Park detention pond. Photo by Ron Flanagan.

- Engage a dedicated, able program champion.
- Establish broad goals, specific objectives, and flexible strategies that can be adapted to avoid land mines, avert problems, and seize opportunities.
- Think holistically. The more comprehensive your program is, the larger your constituencies can be.
- It is important to take a negative mission (such as regulating floodplain use) and convert it into a positive, synergistic mission (such as also providing community parks and open space).
- Find something that is working well and attach your program to it. It might be the Red Cross in one town or the United Way in another, or perhaps the churches or the library or the city council. Every town will have a good starting place.
- Partnerships should be mutually beneficial, and all strategies should be win/win. Learn to listen well to what your partners need, and find ways to deliver it — as long as it does not compromise your base principles.
- Marry opposites for a stronger program. As Dr. Mark Meo at the University of Oklahoma taught us, good public policy happens at the intersection of grassroots citizens and technical experts. It's

109

true with many diverse populations. Engage academics and marry them with common-sense common folk, too, for another example.

- Never, never underestimate the power of the news media. Find ways to inspire them to share your community vision. You are challenged to become a translator of technical jargon into memorable sound bites that motivate humankind.
- Plan to seize any postdisaster window of opportunity. It may be in your town, but you can also take advantage of disasters or trends (such as a jag of interest in green building) elsewhere that capture the public interest. Shamelessly take advantage of the *hazard de jour* and build on it.
- Once you are certain of your long-range goals and principles, dare to invite in your adversaries, listen sincerely, learn from them and seek to convert them into supporters.
- Celebrate success. Always spin to the positive. There are no failures, only lessons learned.
- Find your best management style. We use a jazz-band system we learned from a Tulsa planner named Gerald Wilhite, with light central control and maximum freedom for innovation; shared vision holds it all together and keeps it working in harmony.
- Rejoice in independent successes. Perhaps the best measure of success occurs when people create independent programs that further your mission. The Tulsa motto (perhaps the secret to collaboration success) is "There is no end to what you can accomplish in this world if you don't care who gets the credit."

REFERENCES

City of Tulsa Flood Mitigation Voluntary Acquisition Plan, approved by the Public Works Department, June 21, 1994, and amended April 5, 1999.

City of Tulsa. November 25, 2002. *Multi-Hazard Mitigation Plan,* Consultants Flanagan & Associates, et al. www.rdflanagan.com/tulsa/tulsa_NHM_book.pdf.

City of Tulsa Public Works Department. 1990–91. The City of Tulsa Flood and Stormwater Management Plan, 1990–2005.

City of Tulsa Public Works Department. 1990–95. Annual reports.

Federal Emergency Management Agency. "Background Information for Tulsa, Oklahoma CRS Presentation Feb. 24, 1992."

Federal Emergency Management Agency. "Safe Rooms Take Tulsa by Storm." Mitigation Case Studies.

Federal Interagency Hazard Team. September 4, 1984. *90-Day Post-Flood Recovery Progress Report.*

Federal Interagency Regional Hazard Mitigation Team. June 15, 1984. *Interagency Flood Hazard Mitigation Report*, in response to the May 31, 1984, Disaster Declaration, State of Oklahoma FEMA-709-DR.

Federal Interagency Regional Hazard Mitigation Teams. September 1981. *Flood Hazard Mitigation: Handbook of Common Procedures*.

Flanagan, Ron. October 1991. "Mingo Creek, Tulsa, Oklahoma." *A Casebook for Managing Rivers for Multiple Uses*. U.S. Department of Interior National Park Service, in conjunction with Association of State Floodplain Managers and Association of State Wetland Managers.

French & Associates. August 20, 1993. Floodproofing and Acquisition Program Design Paper. For City of Tulsa, OK.

Garland, Greg. October 21, 1997. "Model Flood Control: Tulsa moves from flood capital to model of control, planning." *The Baton Rouge Advocate*.

Hinshaw, Robert E. 2006. Living with Nature's Extremes: The Life of Gilbert Fowler White. *Tulsa Case Study*, pp. 195–204. Boulder, CO: Johnson Books, a division of Big Earth Publishing.

Kusler, Jon. 1982. "Tulsa, Oklahoma." *Innovation in Local Floodplain Management: A Summary of Community Experience*, Natural Hazards Research and Applications Information Center Special Publication 4, prepared for U.S. Water Resources Council. Boulder, CO.

McLaughlin Water Engineers. June 12, 1984. Flood Hazard Mitigation: Tulsa, Oklahoma, May 26–27, 1984 Flood Disaster, for City of Tulsa, OK.

McLaughlin Water Engineers. July 31, 1984. Revised mitigation plan, for City of Tulsa, OK.

Meo, Mark, Becky Ziebro, and Ann Patton. February 2004. "Tulsa Turnaround: From Disaster to Sustainability." *Natural Hazards Review* 5, 1: 1–9, http://ascelibrary.aip.org.

Mileti, Dennis S., ed. *Disasters by Design: A Reassessment of Natural Hazards in the United States*. Washington, D.C.: Joseph Henry Press.

National Wildlife Federation. 1998. *Higher Ground: A Report on Voluntary Property Buyouts in the Nation's Floodplains*. Vienna, VA.

National Wildlife Federation; Joby Warrick. June–July 1999. "Seeking an End to a Flood of Claims." *National Wildlife Magazine*: 30–33. Vienna, VA.

Oklahoma Water Resources Board. September 1987. "Innovative Tulsa Utility Sets Storm-Safe City as Its Goal." *Oklahoma Water News*.

Patton, Ann. 1975. *Flaws in the Laws and Gaps in the Maps. The Evolution of Floodplain Management in Tulsa, OK 1970–75*. Tulsa, OK.

Patton, Ann. October 7, 1976. "Will Someone Please Make Those Floods Go Away?" *Tulsa Magazine*: 14–17, Metropolitan Tulsa Chamber of Commerce. Tulsa, OK.

Patton, Ann, and George Birt, David Breed, and Randy Kindy. 1976. "In Harm's Way: Flooding in Tulsa: A Case Study in the Creation of Disaster." Presentation for University of Tulsa Symposium on Floodplain Management. Tulsa, OK.

Patton, Ann. 1977. "Why Tulsa Floods." *Tulsa Home and Garden Magazine*: 8–11, 46, and 48. Tulsa, OK.

111

Patton, Ann, and Ron Flanagan. 1980. "Haikey Creek — Flash Floods and the Frontier Ethic." *Intergovernmental Management of Floodplains*, Rutherford H. Platt, ed. Program on Technology, Environment and Man, Monograph #30, Institute of Behavioral Science, University of Colorado. Boulder, CO.

Patton, Ann. May 30, 1984. Memo to Street Commissioner J.D. Metcalfe: "Flood Hazard Mitigation Program." City of Tulsa, OK.

Patton, Ann. May 16–19, 1988. "Flood-Hazard Mitigation in Tulsa, Oklahoma." *Proceedings of the Twelfth Annual Conference of the Association of State Floodplain Managers, Nashville, TN*. Natural Hazards Research and Applications Center Special Publication #19.

Patton, Ann. December 1993. *From Harm's Way: Flood Hazard Mitigation in Tulsa, OK*. City of Tulsa, OK.

Patton, Ann, ed. May 1994. *From Rooftop to River: Tulsa's Approach to Floodplain and Stormwater Management*. City of Tulsa, OK.

Patton, Ann. March 2, 2004. "Together We Can! Public-Private Partnerships for Flood-Hazard Mitigation." Presentation to the Disasters Roundtable Workshop, Reducing Future Flood Losses: The Role of Human Actions; The National Academies. Washington, DC.

Poertner, Herbert G. September 1980, revised 1982. *Stormwater Management in the United States: A Study of Institutional Problems, Solutions and Impacts*. Tulsa Case Study, pages 222–33. U.S. Department of Interior.

Rubin, Claire B., and Mohammad Nahavadian. March 1987. *Details on Frequency of Disaster Incidents for Federally-Declared Disasters*, 1965–1985.

Tulsa Metropolitan Area Planning Commission; Team One, R.D. Flanagan, and Environmental Sciences Corp. 1976. *Flood Information Study*. Tulsa, OK.

Tulsa Partners and City of Tulsa. November 4, 1998. Grant Application to FEMA for Tulsa Project Impact program. Tulsa, OK.

Tulsa Partners and City of Tulsa. March 31, 2002. Tulsa Project: Building a Disaster-Resistant Community. *Report to the Federal Emergency Management Agency on 1999–2002 Activities of Community-Based Partners Working to Reduce Disaster Risk*.

Tulsa Partners, Inc. 2007. *Business Plan and Case Statement: The Millennium Center for Green and Safe Living*. Tulsa, OK.

Tulsa Project Impact; Indian Nations Council of Governments. November 2001. *Community Risk Assessment for City of Tulsa and Tulsa County, OK*. Tulsa, OK.

University of Tulsa Urban Studies Department; Tony Filipovitch, ed. 1977. *Proceedings of the Floodplain Management Symposium*, October 14–16, 1976. Tulsa, OK.

U.S. Army Corps of Engineers, Tulsa District. February 1982. *Tulsa Urban Study Summary Report and 16 Reports on Study Findings*. Tulsa, OK.

U.S. Army Corps of Engineers, Tulsa District. May 1985. *Documentation, Flood of May 27, 1984, Tulsa Oklahoma Metropolitan Area*. Tulsa, OK.

White, Gilbert F., chair. 1966. Task Force on Federal Flood Control Policy. *A Unified National Program for Managing Flood Losses*. Washington, D.C.

White, Gilbert F. 1975. *Flood Hazard in the United States: A Research Assessment*. Institute of Behavioral Science, University of Colorado. Boulder, CO.

Wright, James M, and Jacquelyn L. Monday. 1996. *Addressing Your Community's Flood Problems: A Guide for Elected Officials.* The Association of State Floodplain Managers, Inc., and the Federal Interagency Floodplain Management Task Force. Madison, WI.

Wright Water Engineers, et al. September 1985. *Non-Structural Floodplain Management Hazard Mitigation Measures.* For City of Tulsa and FEMA.

Wright Water Engineers, et al. September 1986. *Guidelines for Inventory and Removal of Structures from Flood Hazard Areas.* For City of Tulsa and FEMA.

Wright Water Engineers, et al. November 14, 1986. *Flood Hazard Mitigation Report, 1986 Floods.* For City of Tulsa.

Young, Terry. February 23, 1984. "City of Tulsa Stormwater and Floodplain Management." Position paper issued during Young's campaign for mayor. Tulsa, OK.

Young, Mayor Terry. July 12, 1984. Transcript of KOTV televised speech on flood mitigation program. Tulsa, OK.

HAZARD MITIGATION IN BERKELEY, CALIFORNIA: PARTNERING FOR COMMUNITY ACTION

Arrietta Chakos

Arrietta Chakos most recently served as assistant city manager in Berkeley, California, where she managed Berkeley's legislative affairs and hazard-mitigation efforts. She coordinated the city's negotiation of a 15-year, multimillion-dollar land use development agreement and directed Berkeley's legislative matters with state and federal legislators and executive agency staff. Managing Berkeley's hazard-mitigation programs included strategic use of six local tax measures (obtained with super-majority voter approval) matched with competitively secured state and federal contributions to reconstruct city and school district facilities. She directed development of California's first municipal hazard mitigation to implement sustainable risk reduction. She has served as a technical adviser on panels for FEMA and for its report to the Congress on mitigation planning; GeoHazards International; the Organization for Economic Cooperation and Development (OECD); the World Bank; the Governor's Office of Emergency Services; the Association of Bay Area Governments; UPMC's Center for Biosecurity; as well as on city commissions and university task forces dealing with seismic safety and hazard-mitigation issues. She is currently completing graduate studies at Harvard University's Kennedy School of Government.

113

INTRODUCTION

Berkeley, California, on the San Francisco Bay's eastern edge, lies directly across the Golden Gate Bridge, in a geologically active region. The precarious Hayward Fault slices through the city's hillside neighborhoods; other regional fault systems, including the San Andreas Fault, pose serious threats. Geologists forecast the region has a 70 percent probability that a major earthquake could strike in the next twenty-five years. The densely populated and developed region is susceptible to an array of natural hazards and has suffered most from wild land fires.

Potential damage from near-field effects of earthquakes, along with other natural hazards including urban/wild land fire, landslide, liquefaction, and creek flooding, is considerable.

Two regional disasters — the 1989 Loma Prieta earthquake and a disastrous 1991 urban wild land fire — moved local leaders to undertake efforts to better protect Berkeley's residents and its built environment. The 1989 earthquake was far enough away from the city to cause only minimal direct damage, but the significant regional damage, social disruption, and casualties coalesced into an alarming call. The subsequent 1991 East Bay Hills fire that hit both Berkeley and neighboring Oakland was more locally devastating — twenty-five people died in the East Bay Hills area and over 3,200 homes were destroyed.

These disasters were reminders that local communities were often unaware of the consequences of development in hazard-prone areas, especially during periods without frequent natural hazard events.[1] The Berkeley area hadn't suffered a significant emergency for some time when the 1989 Loma Prieta earthquake struck, and officials responsible for safeguarding the city's well-being were largely unaware of the existing hazard vulnerability. There were, however, a few local leaders who recognized the disaster safety issues enough to know the need to accurately assess the community's risk and to develop a strategy to best reduce that risk given limited financial resources.

COMMUNITY LEADERSHIP

Berkeley's City Council, at the behest of a council member, Alan Goldfarb, made the first steps toward community sustainability. An urban development expert, Goldfarb had long advocated for more involvement by city officials to respond to and recover from potential disasters. He worked with the regional council of governments, the Association of Bay Area Governments,[2] on a benchmark study about local government's responsibility and liability with respect to disaster readiness. Mr. Goldfarb convinced city council colleagues in July 1989 (a few months before the October 1989 Loma Prieta earthquake) to fund an Office of Emergency Services for Berkeley and pushed for an updated disaster-response plan. These actions prepared the way for the larger community changes that followed.

Once the 1989 and 1991 disasters hit, Berkeley residents saw they had to act on their own behalf to lessen the community's risk. Although the state and federal governments provided assistance for areas struck by disasters, this assistance was by no means enough to fully compensate people and their communities for direct (and indirect) financial and property losses. Berkeley's response in the postdisaster environment was an atypical focus on prevention. Common sense suggested that reducing risk before the next disaster through preparedness programs and strengthening structurally vulnerable buildings were more prudent approaches. Inspired by Goldfarb's leadership, local leaders made community safety and sustainability a common value and developed innovative, practical approaches to effectively reduce hazards' risk with long-term programs to support the viability of these efforts.

STRATEGIC APPROACH TO SURVIVABILITY AND SUSTAINABILITY

The community took an innovative and multilevel approach that included:

1. Developing disaster readiness networks in neighborhoods
2. Strengthening buildings and infrastructure
3. Acquiring state and federal partners to support local initiatives

Disaster Readiness Networks in Neighborhoods

Using disaster readiness as a community theme and collective value, Berkeley strived to become more disaster resilient. The City Council convened advisory commissions, the Disaster Council and the Fire Safety Commission, with appointees tasked to keep community readiness an important policy matter. The dedication and persistence of these groups ensure that hazard mitigation remains a high priority in municipal budget decisions.

Other initial efforts included:

- Activating community residents to form neighborhood disaster-preparedness groups. The city council funded preparedness classes taught by firefighters and medical professionals for those who lived and worked in Berkeley. The proviso was that block captains be assigned to the groups and that neighbors work together in trained teams to help one another in an emergency.
- Thousands of people have participated in the safety trainings and have been encouraged in this work by being awarded special disaster equipment and supplies funded by safety tax dollars. In Berkeley's ten square miles, some fifty disaster-supply locations are established, including one at every public school.
- Annual community disaster drills use electoral precincts as the center of outreach efforts with the exercises similar to get-out-the-vote campaigns with precinct captains and well-organized community contact plans. Such efforts have successfully reached thousands of neighbors in a morning with only a few hundred volunteers, and are kept alive with the neighborhood networks essential to a disaster resilient community.

Strengthening Buildings and Infrastructure

The second aspect of the strategy was to strengthen the built environment to better withstand the kinds of expected regional disasters and to promote long-term sustainability. Local city council and school board members made community safety a priority by bringing attention to the need for local action to prevent future potential damage and sought voter approval of funding measures and developed several incentives programs including:

Between 1992 and 2002, Berkeley's city council took six special tax measures to its voters and got approval for over $362 million to retrofit every public school, every fire station, and essential municipal buildings. As well, a new emergency operations center and new schools were built, a pilot community warning system developed, and a project to install a backup water-supply system to use in case a disaster cuts off the main existing supply begun. Berkeley voters approved all six special taxes for hazard mitigation with a super-majority vote. These taxes generated revenue for the seismic and fire-safety upgrades that took the last decade to complete.

The city government also developed programs without much fanfare to reduce risk in private sector buildings with new approaches to increasing safety that proved successful. The city crafted an innovative tax rebate, another Goldfarb initiative, to fund home-safety improvements. Owners who upgrade their homes are returned a percentage of the improvement costs as a rebate when they sell their house. This has been a successful program over the last fifteen years, and Berkeley's retrofit rate (about 65 percent) is high in comparison with the rest of the region. The program was recently updated with better building standard requirements for more effective improvements. Using other fiscal, technical, and administrative incentives for private sector retrofit, many owners have retrofitted their buildings.

Other popular incentive programs have included a permit-fee waiver for seismic improvements in homes and some unreinforced brick buildings. Grant and loan programs were also aimed to assist eligible low-income seniors, disabled, and other low-income residents make safety improvements to their homes. The city's tool-lending library is a much-needed center for technical assistance where residents borrow tools needed to retrofit homes and confer with helpful staff members. Together, these programs spur community safety in privately owned buildings.

Acquiring State and Federal Partners to Support Local Initiatives

The third step of the strategy was to form partnerships with state and federal agencies to leverage local efforts. Cities are hard-pressed to implement pre-disaster safety programs without the technical and fiscal support of state and federal agencies; though contributions from these sources to local communities are decidedly scarce, they can be used strategically to supplement existing initiatives. Berkeley's officials saw that more intergovernmental affiliations would enhance their opportunity to secure outside technical resources and monies for locally-funded projects. This realization became a crucial turning point for Berkeley, as it sought a working alliance with institutions and agencies that could help guide crucial hazard mitigation planning and decisions.

- Berkeley joined state coalitions of school districts and cities politically active in negotiating state-funded allocations for schools and local government buildings. With dedicated state legislators, the allied communities helped improve California laws on safety guidelines for schools and public buildings. With these

117

amendments, urban communities were able to compete for additional safety funding long out of arm's reach.

- The Federal Emergency Management Agency (FEMA) grew to be a strong ally at the federal level. (See Sidebar 1.)
- Berkeley's city government was in an optimal situation to benefit from a wide range of technical and policy resources in the San Francisco Bay Area. The University of California, Berkeley (UCB) campus provided expert consultation on questions with which city officials grappled. The city regularly worked with technical advisers from the campus on how best to evaluate existing buildings; how to define appropriate levels of risk; and how to make informed decisions about the efficacy of seismic retrofit proposals. The campus's role was quite important as a center with an array of interdisciplinary seismic safety experts — seismologists, geotechnical experts, structural engineers, architects, and public-policy experts. These same experts were typically Berkeley residents and actively worked to promote safety endeavors throughout the community, often giving their time at city council and school board meetings to brief elected representatives. Officials relied on UCB policy and research centers for continued policy and program direction as public sector construction projects proceeded.
- Other governmental agencies joined in — the Governor's Office of Emergency Services, Coastal Region (California OES); the U.S. Geological Survey (USGS); the California Geological Survey (CGS); and the Association of Bay Area Governments (ABAG) — and steered the local governing bodies to consider policy and budget decisions that would be the most cost-effective and provide added measures of disaster protection.

The Bay Area Regional Earthquake Preparedness Project (BAREPP, a division of California OES), and the California Seismic Safety Commission led local presentations on risk, potential regional damage estimates, and mitigation and educated the community on how it might address its seismic hazards.

Public discussion continued at many levels in the community, especially with community volunteers on municipal boards and commissions that directly advised the city council to act responsibly to limit the community's risk exposure. Berkeley's mitigation efforts progressed from its

Sidebar I
Berkeley and FEMA: Working Together

HAZUS

In the mid-1990s, Berkeley officials were invited to serve on an oversight panel in the development of FEMA's HAZUS (Hazards US) loss estimation modeling application that, over time, became a baseline measurement for assessing community risk for many types of natural hazards. Because of this participation, Berkeley was an early HAZUS user and gained further local support for the ongoing hazards work using the findings of early HAZUS studies.

Mitigation Benefit Study

Involvement in the HAZUS project led to more local government advisory support to FEMA from Berkeley for a benchmark study. Working with technical, policy, and practitioner experts, Director Witt's office prepared a report to Congress on the necessity for and benefits of pre-disaster mitigation. The report was instrumental in the establishment of FEMA's Project Impact, the Clinton administration's progressive policy response to the problem of natural disasters and their cascading effects. With seed funding for local governments, FEMA intended to knit together partnerships among cities, the private sector, and citizens to develop indigenous solutions to hazards risk in the nation's most risk-prone areas. This was no one-size-fits-all program. The ingenuity for local disaster problem solving was there with an engaged network of involved stakeholders working together. Berkeley and its activist community were ready for just such an experiment.

Project Impact

FEMA officials knew well how to motivate municipal officials to get the best results from their common efforts. Project Impact's early days were marked by successful national conferences as the project was moved through the country. In December 1999, the annual workshop was held in Washington, D.C., where the cities of Tulsa, Oklahoma, and Berkeley,

continued

119

California, were designated the Project Impact communities of the year. The ripple effect of the agency award reinvigorated Berkeley's commitment to ongoing risk reduction. At the time, California cities had been hit by serious budget cuts and in such circumstances, disaster-safety programs are often the first things to go in deference to the more immediate need for police, fire, and health services.

---

assertive seeking and use of the technical and policy resources at hand. Good ideas came from many sources, and they were scrutinized and shaped by a spirited public process.

JOINING FEMA'S PROJECT IMPACT

The Federal Emergency Management Agency (FEMA) and California Office of Emergency Services (OES) were mentors in Berkeley's early mitigation program implementation. Each agency advised the city through lengthy project planning for retrofit and new construction efforts with funding support to reconstruct schools, fire stations and essential city buildings to modern building code standards for government buildings.

The designation as a Project Impact community gave Berkeley local and state funding leverage that kept critical programs funded for an additional three years.

The City Council allocated local monies, tripling the federal Project Impact funds for the Disaster Resistant Berkeley Program, an umbrella for the city's preparedness, response, and mitigation efforts. This investment signaled continued commitment to safety. Berkeley's interdisciplinary approach called for an involved staff team that leveraged internal technical and policy expertise. This integrative approach strengthened the city's capacity with its overall disaster response system and increased its recovery capacity.

Adding to the existing risk reduction and other incentive programs, Berkeley was spurred to deal with other technical issues with Project Impact state and federal partners including:

- Ongoing community organizing and safety programs — Community engagement efforts were further energized by Project Impact AmeriCorps participants made up of a thirteen-person team that worked for three months on neighborhood group and business preparedness. The team's ambitious schedule and outreach program

---

Sidebar 2
Public/Private Partnership

The review of hazardous apartment buildings funded with FEMA monies was a unique collaboration among the city, the Federal Emergency Management Agency (FEMA), the Lawrence Berkeley National Laboratory (LBNL), and the Earthquake Engineering Research Institute (EERI). An inventory and risk analysis of most city buildings had been compiled some years before, but one class of buildings — soft story, multi-unit apartment buildings — had yet to be inventoried. In the 1995 Northridge earthquake, soft-story apartment buildings sustained serious damage, causing numerous deaths. City leaders wanted a realistic assessment of the threat posed by these buildings and did not want to ignore a potential residential-risk problem. In an unusual collaboration, senior EERI structural engineers and UC Berkeley engineering graduate students worked in teams to survey 150 buildings in fall 2001, gathering construction details on the most at risk buildings. LBNL engineers, on a parallel project, did further research on this building type with local senior citizen residents, owners of a particularly vulnerable building. LBNL contributed to the development of sensing devices and conducted tests to help design effective retrofit options with additional grant funding from Hewlett Packard.

---

involved thirty-five small businesses and fifty neighborhood block preparedness groups to renew safety plans and training. Project Impact's $335,000 seed funding was matched by three additional grants that supplemented by $60,000 grants from VISA USA and the State Department of Insurance for public information and home retrofit projects for low-income and senior homeowners.

- The community outreach activities revitalized the community with mitigation and preparedness work and built a more solid connection with business owners. (See Sidebar 2.) This, in turn, strengthened voter support for a 2002 safety tax measure in that funded more neighborhood network building through the emergency equipment and supply program. Without Project Impact support to fund compelling public-information materials and outreach events, it would be difficult to know if continuing support would have been sustained.

- Hazards Mapping Project — The Hazards Mapping Project was completed with help from Project Impact connections with USGS and the CGS scientists with a compilation of the geologic hazards in Berkeley's immediate environs. A multi-agency team worked two years to assemble private and public-sector data and to map areas vulnerable to landslide, liquefaction, and ground-shaking hazards. The completed hazard maps delineate risk zones and were building blocks for more thorough loss-estimation scenarios that better refined local risk analysis.
- Land Use Plan — With the federal imprimatur, the city mustered resources to take a more comprehensive approach to risk reduction and to solidify a cohesive program embedded in both the city's general and mitigation plans. Mitigation policy adoptions in these long-term land use planning documents would weave needed safety policies into the community's daily life. Berkeley's updated land-use master plan was published in 2002[3] after 52 public meetings over a three-year period. The final product included newly crafted disaster safety policies and mitigation goals in the Plan's housing, transportation, and safety elements. City staff and the Planning Commission ensured that seismic safety was emphasized as a primary goal in the policy framework. Inclusion of these policies is a critical factor in the city's development and land use planning, providing legal and regulatory authority for sustainable mitigation programs.
- Community Hazard Mitigation Plan — Publication of the general plan was a necessary step in the development of the city's comprehensive mitigation plan. This was followed by a community summit in December 2003 with 120 representatives from regional agencies and utility providers, and nonprofit agencies, along with the city, campus, and local school district. The summit focused community participants to come to consensus about how to affect disaster resilience and to identify future opportunities for working together. As a FEMA-designated Disaster Resistant University, UCB was also engaged in aggressive mitigation efforts with an ongoing $1 billion retrofit initiative. Berkeley's city council adopted California's first comprehensive Disaster Mitigation Plan in June 2004,[4] and recently updated that initial plan. The Mitigation Plan signaled the community's commitment to safety, signifying the culmination of many years' investment in sustainability.

Sidebar 3
Why Project Impact Worked in the Bay Area

Communities face new challenges in the midst of daunting climate disruption threats. Given potentially insurmountable conditions, effective change, and strong community and government partnerships will be necessary elements of a successful response strategy. The Project Impact model worked because it was a healthy melding of local and federal interests. Communities were valued as genuine partners that brought resources, skills, and mitigation solutions to the table; FEMA's sponsorship without intrusion brought support that further encouraged local capacity.

These outcomes from the Project Impact partnerships demonstrate what a powerful model FEMA promulgated. It literally sowed the seeds of community change and empowerment in hundreds of communities. These cities and regions also formed useful networks that moved forward collective efforts through the sharing of best practices, formation of colleague alliances, and active advocacy for hazard mitigation at many levels.

The program embodies an adaptive leadership model[5] as defined by Harvard University's Ronald Heifetz that defines how to mobilize communities to tackle the problems before them, activating the widest range of site-and problem-appropriate solutions. Project Impact served a crucial function as it used the best of its federal powers to enable widespread community transformation. The San Francisco Bay Area communities involved in the program — Oakland, San Leandro, Napa, Berkeley — modeled the community-protecting behaviors we need to see more of in our world. With neighborhood and business leaders as equal partners, the work to create disaster resistant and sustainable cities was equitably and successfully begun.

CONCLUSION

After two regional disasters struck, community leaders and public officials in Berkeley, California, acted to reduce risk in the city. The investment in long-term sustainability came from many fronts and was a community-wide effort to keep disaster readiness and safety initiatives front-burner issues over an 18-year period. Though these efforts were initiated locally in response to severe hazard events, the city found an ally in the Federal Emergency Management Agency and its Project Impact initiative. Project Impact became a regional umbrella in the San Francisco Bay Area that connected many independent risk-reduction efforts in the hazard-prone area.

1. A. Chakos, P. Schulz, and L. T. Tobin, "Making It Work in Berkeley: Investing in Community Sustainability," *Natural Hazards Review* 3, no. 2 (May 1, 2002), American Society of Civil Engineers Reston, VA USA.
2. Association of Bay Area Governments, *The Liability of Local Governments for Earthquake Hazards and Losses: Report: A Guide to the Law and its Impacts in the States of Alaska, California, Utah and Washington;* 52 p.; February 1989.
3. City of Berkeley, *California General Plan: A Guide for Public Decision-Making,* adopted April, 2002, Safety Element Revised 2003.
4. City of Berkeley, *California Disaster Mitigation Plan,* adopted April 28, 2004.
5. Ronald Heifetz, *Leadership without Easy Answers* (Cambridge, MA, and London, England: The Belknap Press of Harvard University Press, 1994), 128.

5

County/Regional-Based Hazard-Mitigation Case Studies

INTRODUCTION

Large disasters such as hurricanes, droughts, and floods often cross jurisdictional and natural boundaries. These events wreak damage and destruction regardless of local, state, or national borders. Scientists are predicting that because of certain climate change impacts such as sea-level rise and the warming of the oceans, we can expect larger, more widespread disasters in the future. In order to reduce the impact of these types of events in the future, it will be necessary for community leaders to look past their local jurisdiction to county, regional, and, in some cases, international-based approaches.

This chapter presents three case studies that clearly illustrate how multiple jurisdictions and communities can come together to address a shared risk. The first case study, concerning the Living River Flood Management project in the Napa (CA) River Valley, highlights several elements critical to the success of a regional approach to risk reduction, such as a county-wide planning process, involvement of the private sector, detailed knowledge of the risk and potential mitigation measures, and participation by the entire population of the county in making the plan a reality.

The second case study examines how the International Flood Mitigation Initiative (IFMI) brought together government officials, scientists, advocates, environmentalists, businesspeople, and everyday people from Minnesota, North Dakota, and Manitoba, Canada, to build a consensus around a series of actions designed to reduce flood impacts in the Red River Basin.

The final case study of Seattle Project Impact details how Seattle leveraged seed money from FEMA's Project Impact initiative to better understand their earthquake risk and to design and implement three local mitigation programs to protect local home owners, schoolchildren, and small businesses, which were then implemented across the region.

LIVING RIVER:
THE NAPA VALLEY FLOOD MANAGEMENT PLAN

Dave Dickson

David Dickson is currently a senior consultant to MIG, Inc., a California-based planning and design firm. Mr. Dickson consulted with the Federal Emergency Management Agency (FEMA), the Environmental Protection Agency (EPA), the Army Corps of Engineers, the University of California, and George Washington University in the areas of watershed management, restoration, disaster management, and financial planning. His public agency work has included positions as chief financial officer of the Napa County Flood Control and Water Conservation District and Community Development Director for the Napa County Administrator's Office. He was project manager for Napa Valley's "Living River'" Flood Management Plan — a comprehensive watershed-wide plan for flood damage reduction, river and watershed restoration, and economic revitalization in the city of Napa. He was the architect and manager of the Community Coalition planning process and the financing plan of this county-wide effort, which now totals over $500 million in public investment. He holds a B.A. in political science from San Diego State University and has completed master's-level course work at the Public Finance Institute, University of California, Davis.

From all indications, we are entering into an era of natural disasters. Even though the causes of this change are global, the effects will be very local, affecting each of the communities we live in. In the community where I reside, California's Napa Valley, we have already had a preview of the devastation that climate change will bring. For our entire history, we have been overcome by major floods that destroy our community, taking away our lives, property, and peace of mind.

126

Unfortunately, flood-induced disasters like those that Napa faces will only become more frequent in California and elsewhere in the years ahead. Scientists who study weather patterns predict that the Bay Area, in particular, will be slammed with more extreme storms bringing more intense rainfall in the coming years. This is supported by research conducted by the National Oceanic and Atmospheric Administration (NOAA), the U.S. Geological Survey (USGS) and other scientists at the recent California Climate Change Conference, sponsored by the California Energy Commission and the California Environmental Protection Agency.

These same scientists predict that climate change in California will cause three troubling outcomes that will ultimately threaten the health and safety of every community. The first result is an increase in severe "Pineapple Express" storms from the Hawaii island area. These storms carry intense amounts of warm rain that will lead to more flooding. The next major effect will be further loss of the Sierra snowpack as temperatures increase, leading to drinking water shortages. Finally, climate change will result in drier, warmer weather inland, leading to more wildfires.

How does this affect flood protection? Throughout California, levees, dams, flood-control channels, and bypass channels are being forced to manage water flows for which they were not designed. Even the most forward-thinking 1950s estimates of peak flood flows, such as those engineers designed for on San Lorenzo Creek, are now being shown to be at least 50 percent below what will now flow from the hills during each Pineapple Express storm.

This means that more and more communities will need to address the threat of flooding, or risk the economic deterioration experienced by Napa over its history due to major, frequent floods. As we know, a community that is not economically healthy is not healthy. The community's heart is under attack, as economic problems cause social problems and put strain on almost every member of the community. This is the human aspect of what's at stake in flood protection.

Yet because of the huge cost of multi-objective flood protection mitigation, only a Napa-like planning and Community Coalition process is likely to result in the action needed to upgrade the flood-protection infrastructure of these communities.

California, in particular, has strict laws requiring voter approval of any new special taxes for flood control. A two-thirds "super majority" is required. It used to be that communities relied on the U.S. Army Corps of Engineers to pay for 80 percent or more of major infrastructure improvements, but those days are gone. The recent Water Resources Development

Act authorized an additional 180 projects around the country to receive federal help. Yet, if recent history is any measure, federal appropriations for these projects will be a long time in coming, if they come at all. This will become more and more the case as climate change becomes a reality, forcing the federal government to transfer its limited dollars to the "crisis of the year," such as Hurricane Katrina, leaving just enough funds to spread around to keep all of the other urgent projects going forward, if barely.

The Napa River project provides a case study of how a community has come to terms with its river and its flooding problem in a successful way. In this article, I will tell the story about the genesis of the Community Coalition Planning process that secured the agreement and political support needed to pass a sales tax to raise the local share of what has turned out to be a $500 million investment in "Living River" flood protection throughout the Napa Valley.

The project has been under construction for ten years now. What has been accomplished? What still needs to be done? The second section provides a project update, including "the good, the bad, and the ugly" of executing the largest public works project in the history of Napa. The final section concludes with some "lessons learned," and outlines the elements that need to be in place for a disaster-prevention project of this size and complexity to be successful.

The Napa River is a thread that runs through the Napa Valley. Starting from its headwaters on top of Mt. St. Helena, the river levels out and meets up with the San Francisco Bay Estuary in the city of Napa, the major urban center of the Napa Valley. Given its position on the river, it is not so remarkable that the city sits where it does. The city is centered where the river meets Napa Creek and then turns back on itself in what locals call the "Oxbow," making it the furthermost navigable point on the Napa River Estuary. The tides come in and out up to this point, about a third of the way up the 55-mile length of the river, which runs from Mt. St. Helena to the San Francisco Bay.

The Napa Valley community has had a love-hate relationship with its river since the area was settled in the mid-1800s. For decades, the river has provided fresh water for the Valley's many farms and vineyards, which still comprise its main industry to this day. Beautiful and idyllic, the river has also provided a home for fish and wildlife and a place for people to relax and play.

However, when it starts to rain, the river takes on a much more dangerous and threatening character. It floods over its banks, causing damage and loss wherever it flows. Unfortunately, this happens all too often. Napa is one of the most flood-prone communities in California, even though we have a total population of only 126,000 people. Since 1862, Napa Valley has endured 27 major floods. This can be devastating in the heart of downtown Napa, where the river can carry only 20,000 cubic feet per second. In 1986, in the largest flood in Napa's recorded history, close to twice that volume overflowed the riverbanks. This "100-year" flow also inundated the region in 1995 and 2005.

As with most rivers around the world, each time a serious flood happens, the community goes into crisis mode. It is best described by that knot in your stomach when you know people are being traumatized in your community, especially the elderly and more vulnerable, who invariably end up living in the floodplain because that is where the cheapest housing is. One is also thankful at these times for the emergency response system — the fire departments and human-service system, the shelters, the police, the water rescuers, the volunteers who bring food to the shelters, and the innkeepers who provide rooms to the evacuees. It is government at its best! Unfortunately, it was not that way in the 1986 flood, because there had not been a major flood in about fifteen years, and people forget about floods quickly.

It is even worse in the small towns of Yountville and St. Helena, halfway up the Napa Valley, where one third of the town's housing stock is in the mobile home parks, which were built in the 1960s, before floodplain regulations. These mobile home parks have flooded regularly. The people in these parks are the ones that I think about during high water.

Historically, floods have not been the only problems connected to the Napa River. Fifty years ago, slaughterhouses, tanning factories, sanitation districts, and oil companies discharged their wastes directly into the Napa River. The river was diked and leveed, and industrial buildings and residences were built right on top of the natural floodplain terraces of the river.

But the tide has changed for the Napa, and today the river is arguably one of the most important waterways in the nation. A dedicated and diverse community of activists and agencies that fought to resurrect it has not only improved its water quality and secured thousands of acres of wildlife habitat along its banks but has created an important model that redefined America's approach to flood control.

Our community has tried to fashion a solution to the major flooding for its entire history. Since the 1960s, no fewer than four U.S. Army Corps of Engineers proposals have been presented, voted on, and rejected. The projects proposed in the 1960s, the 1970s, and then again in 1995 just did not address the needs of the Napa community. They did not protect its migrating

FIGURE 5.1 Napa, CA, February 14, 2006 — This California resident raised their home ten feet to prevent, or mitigate, flooding. Photo by Adam DuBrowa.

fish, riparian zone, and wetlands and did not protect it from floods or reduce the potential damage floods could cause to Napa's 7,000 downtown structures, including its civic center, not to mention the lives of its citizens.

Then a remarkable "coming together" occurred around flood control, which voter surveys said was the number one issue facing the community. Over a 30-month Community Coalition process, the community's business leaders, environmentalists, government officials, mobile home owners, neighborhoods, fishermen, canoers, Red Cross workers, gadflies, and others participated and coalesced around the concept of a "Living River" flood-protection and restoration plan for the Napa River. On March 3, 1998, the voters weighed in with the required two-thirds majority to raise taxes in Napa County in order to implement the Living River Flood Protection and Estuary Restoration Plan.

I had the fortunate opportunity and privilege to manage the planning process and build the community-based structure needed to bring about compromises and achieve this community consensus. I was the process architect and manager. It helps that I am a self-confessed consensus junky. I have lived and worked in the Napa Valley community for over 30 years, and I had the networks, relationships, and understanding of the parochial and esoteric political sand traps that exist here.

The first thing our community demanded was that the U.S. Army Corps of Engineers change its relationship with the community. The community wanted to take control of their county's government to make it work for the community. The Corps needed to agree to come out of its offices and mix it up with the community. It needed to hear and listen to us. The San Francisco Bay environmental community had to be embraced and accommodated, because the 425-square-mile Napa River watershed is the last undammed tributary flowing into the San Francisco Bay Estuary. It is also a critical salmon and steelhead habitat and home to special status species, including California freshwater shrimp, salt marsh harvest mouse, and California Clapper rails. The regulatory agencies made it clear early in the process that they would not permit a typical Corps approach of encasing the river in concrete.

With the help of Senator Barbara Boxer, we got the Corps to the table, agreeing to use the congressional planning appropriation of $1 million to focus on the local Community Coalition process. The Corps needed to "trust the process," but it was a new experience for them in many ways.

A Coalition Steering Committee was formed, composed of local elected officials and the presidents of the Friends of the Napa River, the Napa Valley Economic Development Corporation, and the Wine Institute. The first thing the committee did was develop a set of goals:

1. Protection from the 100-year flood
2. A living, vital Napa River
3. Economic revitalization
4. A cost that the citizens could support
5. Retaining our valuable federal project authorization (50 percent funding)
6. Watershed wide planning and a solutions-integrated "system"

In essence, they wanted it all. In order to achieve a two-thirds vote on a tax increase, every influential sector of the community had to be satisfied — in fact, excited — about transforming a floodplain. So the goals were presented to a coalition of 27 local stakeholder organizations to see if they would commit to a process to develop a flood plan addressing all of the goals. If, in the end, they could not commit, well, at least we had given one last concerted community effort.

Everyone warily agreed to sit at the table and assist in "resourcing" the effort: the Corps as well as the 27 government agencies with jurisdiction over the Napa River and any development within its sphere of influence. Over 24 months, there were eight town-hall-type meetings involving 200 to 250 of Napa's finest minds, who actively participated to conceive a plan, check its constructability and science, and determine its financial feasibility. These meetings became a celebration of progress. Over the first six months, the theme of the Living River became a rallying point, a point of guiding light against which any idea would be tested to determine if it contributed to it or threatened the achievement of that goal.

The hard technical work took place in a continuous process to support the coalition's direction. The coalition hammered out financing plans, urban-design concepts and standards, and definitions of the Living River based in science. The community learned about things like dissolved oxygen levels, continuous fish and wildlife riparian corridors, geomorphically stable channels, and a river system's natural width-to-depth ratio. We were told how we fit in the big picture by the likes of Luna Leopold, the son of Aldo Leopold, the great environmentalist. Luna was in his late seventies at the time and is considered the father of modern river geomorphology.

The old timers of Napa have always believed that no flood-control solution was possible because of the tidal action in the Napa River. The scientists sat with them and talked these things out. The scientists had to demonstrate with computer models how the tides interact with the flood flows, and how the Living River Plan accommodated both flows to protect the city from flood damage. We learned right away, of course, that you cannot control floods. You plan for living with them. The lessons of 1993 on the Mississippi River and the Galloway Report were vital to the coalition.

The four technical committees were organized according to different focus areas: Living River, Up-Valley watershed management, urban design, and finance and regulatory issues. Each was made up of a cross-section of paid staff, government staff, hired consultants, the Corps of Engineers, and local citizens with special capabilities such as landscape architecture, natural-resource management, and political organizing. The committees met in the same auditorium each Friday for six months, preparing details to present to the larger Community Coalition.

The community held a celebration of achievement at Chardonnay Hall at the fairgrounds with over 200 coalition participants in June of 1996, when the concept was developed enough to pronounce it a plan. Then began a one-year period of verification, to see if the details supported a plan that could actually be implemented.

After two years of relentless and intense research and negotiations, the Corps, 27 other governmental agencies, and 25 local nongovernmental organizations hammered out a revolutionary "Living River" plan. Where the Corps had proposed floodwalls and levees, the Coalition proposed terraced marshes and broad wetlands. Where the Corps had proposed dredging the river deeper to allow it to carry more water faster, the Coalition proposed making it wider, by returning much of its floodplain.

The plan had stiff requirements. We wanted to reconnect the river to its natural floodplain and maintain the natural depth-to-width ratio of the river. We wanted to restore historical tidal wetlands and implement watershed-management practices to maintain the natural riparian corridors along the river and tributaries. We needed to clean up contaminated river-adjacent properties, replace eight bridges that now act as dams during high flows, relocate, purchase, or elevate 150 homes, businesses, and mobile homes that were in the floodplain, and purchase over 900 acres of river adjacent agricultural lands.

Original estimates for the plan totaled $250 million. About $100 million was to come from the federal government and state environmental restoration grants and highway bridge funding. $150 million was to come from local taxpayers and the tourists who visit Napa Valley. A half-percent increase in the local sales tax taps the tourists, who pay about one third of the local sales tax. This was a very appealing feature of the finance plan to the citizens. Other tax-increase proposals were soundly rejected in community surveys conducted under the direction of the Community Coalition.

The Coalition said the tax must expire after 20 years, and two citizen oversight committees were required in the tax measure to scrutinize expenditures and oversee the technical aspects of project implementation.

Professional public opinion surveys were conducted. By March of 1997, the plan was verified to a point that we knew the voters would support it, the Corps of Engineers could participate in an environmentally restorative program of flood management instead of flood control, and the environmentalists would compromise and ultimately support the tax increase and actively campaign for the effort.

The community coalition process itself became the campaign. All 27 organizations at the table either supported or were silent during the campaign. A well-financed public-issue campaign was bankrolled by individual contributions, investments by several large wineries who wanted to bring the city of Napa into a more intimate relationship with the wine industry and Up-Valley ambience, and by the environmental and business communities. Groups that are usually at odds came together around the flood problems of Napa.

All five cities of the Napa Valley and the county agreed in a Joint Powers written agreement on how the tax proceeds would be equitably shared to address flood protection on a watershed-wide basis.

On March 3, 1998, 23,000 Napa County voters turned out in a special election to vote on Measure A, the flood-control measure. Only one issue was on the ballet. It was a very high turnout for an election like this. At the end of the evening, the community celebrated victory, with a 300-vote margin. At the end, every participant in the process felt that his or her efforts had made a difference. It was a very sweet victory.

Project Update — A Community Lives Through It

After ten years, the Napa River Living River Flood Protection and Estuary Restoration Project is about 75 percent complete. It has ushered in a new era for the city of Napa and a major transformation of the city's southern entrance and downtown. Old levees have been removed or breached, creating more than 1,000 acres of new wetlands. Five new bridges that used to act as dams during flood flows have been reconstructed. The city has managed to survive the major community disruption that is the result of such a massive undertaking. Costs have almost doubled over the original estimates for the project, but fortunately higher-than-expected proceeds from the half-cent sales tax and State of California bonds for flood control have managed to keep the local expenditure side of the equation in balance. The Federal Corps of Engineers' funding, however, has lagged, therefore postponing flood protection.

South Wetlands Opportunity Area

At the entrance to Napa in the southern reaches of the project's seven-mile span, the first phase of the project, known as the South Wetlands Opportunity Area, was completed in 2001. Levees were removed and breached to allow the tides to restore a marsh plain of about 1,000 acres, which floods twice daily during high tides. A floodplain that is at a slightly higher elevation and is inundated once every two years on average was also created, essentially giving the river back its bank-side sponges. The marsh and floodplain has combined with other terracing and grading along the river to help lower downtown water surface elevations by several feet during flood events.

The Corps of Engineers and the California Department of Fish and Game, who have been monitoring fish in the restored floodplain for the past six years, are finding that native fish seem to be drawn to the new marsh and floodplain. Shorebirds can be seen in abundance probing in the new mudflats while ducks fly overhead. At a recent Bay Area science conference, the Corps said that the floodplain areas have increased rearing habitat for fish.

Heather Stanton, the Project Manager at the Napa County Flood Control District, emphasizes what a unique opportunity the community had. "The Project is very land-intensive in terms of restoration. The whole theory is to return the river to its floodplain where possible. That would be a constraint for some cities."

Oil Company Road

Another physically intensive phase of the project involved purchasing, demolishing, and relocating 33 buildings and warehouses, including nine residential units and 53 mobile homes, as well as relocating the Napa Valley "wine train" tracks and concessionaire building. According to Stanton, while residents were given relocation assistance and most moved willingly, some of the industries had a difficult time finding new sites. Pulling back the riverbanks — where industries on Oil Company Road were located before tankers started bringing heating oil and gasoline to Napa in the 1850s — also meant surprises. The city inherited 11 properties contaminated with petroleum, and despite the regional water board's promise to remediate the sites, the Flood Control District ended up paying for the removal of 170,000 cubic yards of contaminated soil. Ultimately the oil companies who polluted the land were forced to help bear the cost.

This phase and expensive land acquisitions in the north, along with the discovery of unforeseen underground utility, sewer, and water lines in the excavation area, were the primary causes of the 100-percent increase in the local share of the project budget.

Urban Riverfront

Across the river from the formerly contaminated sites — now new mudflats and floodplain — Army Corps flood walls have been completed and architecturally upgraded with local city funds. The walls double as protection for the historic Hatt Building and the 1884 granary, now called the Napa Mill Complex, designed and financed by local developer Harry Price to include an upscale inn, general store, several restaurants, and a pie shop. On top of the flood walls is a pedestrian promenade where walkers, bicyclists, and tourists can gaze out over the river and watch the bird life in the mudflats and marsh.

The more traditional, urban downtown stretch of the project was a key area of compromise during the Community Coalition process. The environmentalists wanted natural river on both riverbanks but agreed to allow fortification of the bank on the city side through the urban downtown reach.

The City of Napa, along with consultants, is guiding the project design, including the reconstruction of four new downtown bridges in

135

accordance with a new River Parkway Master Plan. The main idea is to turn adjacent development to face the river where possible and to treat the entire river corridor as open space. The city has made a major effort to ensure that the design for the river edge adds value to adjacent properties. Four bridges have been built and were designed to reflect the aesthetic of the floodwall promenade.

Oxbow Bypass
The wet-dry bypass for the large, horseshoe-shaped Oxbow turn in the river is a key feature of the flood-protection strategy. The bypass is a 900-foot-long, 280-foot-wide channel that will divert flood flows away from the Oxbow during large storm events and help speed the water downstream. Until the bypass is constructed, the level of flood protection remains less than needed and flooding can occur.

It is here, at the Oxbow Bypass, where the Community Coalition and Army Corps design process has gotten bogged down.

The latest incarnation of the bypass, designed with input from an advisory panel, is a seasonally dry channel that will act as a metaphor for the entire Napa Valley. The sides will be planted with native trees and grasses to resemble the Valley hills, while the bottom will replicate the Valley floor with a grid system of boulders, a lawn (so that flood flows are not a concern), and finally, a small "low-flow" channel, which is a sort of analogue to the Napa River.

How many of the trees and other design features proposed by the community the Corps will accept remains a question. According to city staff, Friends of the Napa River members, and other Community Coalition members, when it comes to the Corps and community-based design, "the devil is in the details." It is hard to get these big federal entities to get down to the level of detail that the community is interested in. Some members of the original Community Coalition design team worry that the softer approach promised by the Corps seems to be getting harder. They are worried that areas where banks were to be stabilized predominately with plants will now be covered by more rock than was originally planned.

For issues like this, the Coalition can count on the Technical Advisory Panel established in the sales tax ordinance to act as a bulwark against too much mission creep on the part of the Army Corps.

To the north of the Oxbow is one of the Coalition's most prized achievements: 12 acres of riparian forest tucked into the Oxbow, acquired with $3 million from the State of California's urban Streams Restoration program. The Coalition has implemented a plan to preserve the Oxbow's native trees while removing invasive plants, adding some wetlands, and restoring old sloughs.

The Response of Local Elected Officials

Overall, local officials in Napa are pleased with the results of the Project for the City of Napa, although they are frustrated by the slow pace of funding from the Army Corps of Engineers. Maintaining political support and enthusiasm is critical to satisfactory completion of a long-term public works project.

One especially enthusiastic supporter of the project is the Mayor of the City of Napa, Jill Techel:

> This project looked good on paper but as it has evolved it is even better. The first results were the reclaimed wetlands. The wildlife that has returned and grown is amazing. It really is something to see the birds and especially the new chicks running along the shoreline and getting ready to fly during springtime. The new trails are so appreciated by the local community. We had placed our back to the river and now we are opening it up and it is every bit the treasure we thought it would be. The trade-offs are key. We had multiple goals to meet them there was and continues to be give-and-take. We had historic structure issues, resource agency issues, safety issues, economic future issues. However, as long as everyone takes a step back and considers what is good for the City in the long-term, compromises can be made and the project can move forward.

The Napa County Flood Control and Water Conservation District has also been a key, local player and observer of the project. Bill Dodd, who is Chair of the District, summed up the benefits of the project succinctly:

> The project has increased the safety of Napa residents, spurred economic development, cleaned up acres of contaminated riverfront, and restored important habitats. Since groundbreaking in the summer of 2000, the project has reached multiple major milestones, including restoration of over 1,000 acres of historic wetland in the Napa River estuary, construction of five bridge replacements, and cleanup of 11 acres of riverbank contaminated by petroleum spills. The severity of flooding has already been reduced by the partially completed project. This was clearly shown during the flood of December 31, 2005. The restored wetlands, mudflats and tidal terraces have stimulated huge growth in the bird count and expansion of riparian and riverine habitat. Partially complete river trails are highly popular in the community.

**Put to the Test: Half-Completed Flood Project Survives
Napa's Third 100-year Flood in 30 Years**
As Bill Dodd observed, even at partial completion, the project has mitigated the severity of flood events dramatically. Nowhere was this shown more effectively than on New Year's Eve 2005, when the partially completed project was inundated with the third 100-year flood since 1986. A big storm hit the region that night. Nearly ten inches of rain fell within a 24-hour period, causing the river to rise to 23 feet and overflow its banks, flooding the entire region again, including downtown Napa. The question for everyone involved in the project was: How will we fare?

The *Vellejo Times Herald* measured the situation a few weeks later, in its January 23, 2006 issue: "Despite a thorough drenching that flooded Napa's downtown and caused an estimated $70 million damages to the city and $115 million worth of losses countywide, the half-completed Napa River flood control project still proved its worth." According to Napa County's emergency services manager, Neal O'Haire, the project worked in part to draw the water south, away from the Napa Valley and into the marsh and wetlands (reported in an article in the February, 1, 2006, issue of www.bohemian.com). For Jill Techel, Mayor of the City of Napa, the partial success of the project was clear: "The good news, if there is good news when you flood, is that because of the work that has been done the water receded in less than 24 hours. In the past it has stayed for three days."

Put to the test, the project proved its worth from a flood-protection perspective even before completion. Unfortunately, though, a flood-control project is sort of like re-roofing your house: even with 75 percent completed, you still get wet.

Funding: Good News and Bad News
The most successful aspects of the project are the transformation of the City of Napa and the surrounding environment, flood protection, and economic development that has and is occurring. The key objectives of the project are being achieved. A less successful aspect has been the rising costs and cost overruns required to achieve implementation.

Fortunately, several additional sources of local and government funding have allowed the project to proceed with cost overruns of almost 100 percent in the cost of land acquisition, relocation, and utility relocations. It has been somewhat disappointing that the spirit of the Community Coalition has not carried over to property negotiations, where some "opportunism" has occurred. The City of Napa portion of the project cost is now about $350 million and another $175 million has been spent or is scheduled to be spent in the upper Napa Valley communities.

Funding for the local share items of the project has kept pace with cost increases through the achievement of "layered funding" from multiple federal, state, and local funding that results from the advantages of a multi-objective project that includes elements beyond flood control. These sources include:

Table 1: Multiple Benefits Equals Multiple Funding Sources

U.S. Army Corps of Engineers	$130 million
Grants/Loans: Environmental Restoration	
California Coastal Conservancy	$2 million
State Lands Commission	$2 million
CALFED	$5 million
DWR Urban Streams Restoration	$1 million
California River Parkways Grant	$4 million
Clean Water Act: SRF, 2 percent Loan	($65 million)
FEMA Hazard Mitigation Grant Program	
Napa Creek Home Relocations	$1 million
Mobile Home Park Protection	$7.5 million
Countywide Home Elevations	$4.5 million
Floodway Buyout	$5 million
Yountville Floodwall/St. Helena Planning	$8 million
Federal/State Highway Funds	
3rd Street Bridge	$9 million
1st Street Bridge	$2 million
Maxwell Bridge	$22 million
State Road Subventions	$75 million
Local Sales Tax Increase	$240 million
Total Project Cost (includes Maintenance Trust Fund)	$520 million

Unfortunately, the Corps of Engineers funding has not kept up with the local financing. Under the cost-sharing agreement with the Army Corps, the local community is responsible for the cost of lands, easements, rights-of-way, relocation of utilities, structures, and railroads (except railroad bridges). Almost 100 percent of the local elements are completed. The Corps' portions — including all excavation, setback levees, terrace grading, in-channel work, and railroad bridges — are in various states of completion, due to funding shortfalls.

The reasons for the shortfalls are complicated, but part of it is due to the long process of securing money from the federal budget every year, as reported in the *Vallejo Times Herald* on January 23, 2006, a few weeks after the 2005 New Year's Eve flood: "Every year the Army Corps, the President and Congress go through a complex bartering system to reach an agreement on funding for the Federal share of the Napa project. In 2005, the Army Corps of Engineers estimated the project needed $34 million. It received $12 million." The next year was not much better. The Army Corps of Engineers estimated that it needs $24 million for the project in 2006. The president only budgeted $6 million. Ultimately, the project received $14 million, according to the July 18, 2007, issue of the *Napa Valley Register*, which is still significantly less than what the Corps needs to complete its project tasks.

Given the cost of flooding in terms of property, cleanup, and lives, the federal government may be acting pennywise and pound-foolish. As reported in the January 23, 2006, issue of the *Vallejo Times Herald*, using a slightly different metaphor, Barry Martin, spokesperson for the Napa County Flood Control and Water Conversation District observed, "The government has probably paid more than the cost of the whole project in just damage claims and cleanup over the course of the last three floods. Rather than treating the symptoms, you're treating the disease when you finish the whole project."

Supporting Martin's observation is the fact that since the project's inception in 2000, the region has saved an estimated $1 billion in flood-induced damages, indicating that an ounce of prevention really is worth a pound of flood-induced damage claims.

Economic Development

The story of the economic benefits of the project for the region is perhaps best told by someone who was "on the ground" at the time, Steve Kokotas, former executive director of the Napa Economic Development Corporation:

The project has fostered a major economic boom in downtown Napa and throughout the Napa Valley. By simultaneously mitigating the risk of flood and improving the natural beauty of the riverfront, the city was able to attract developers and other investors interested in transforming downtown Napa into the world-class destination it deserves to be. The last six years have seen the construction of new hotels, spas, resorts, arts centers, wine-tasting rooms, and other commercial, arts, and entertainment development along the riverfront. More than a dozen vineyard owners opened up downtown tasting facilities that increase their products' visibility while reducing the negative impact of heavy traffic and tourism in the Napa countryside. All of these new amenities, along with the new natural and recreational opportunities provided by the restored river, entice visitors to stay longer and contribute more to the local economy. They have also created many job opportunities. In short, by protecting the downtown from major flooding, the project turned the town's biggest liability, the river, into its greatest asset.

The project has significantly improved the urban landscape of Napa. The most noticeable aspect is the influx of private investment and development in the region, surpassing $560 million since 1999, according to the Napa Community Redevelopment Agency. The new $70 million COPIA Center for Wine, Food, and the Arts was an important early anchor of the downtown riverfront revitalization, representing the first time the city has been able to enjoy the benefits of the Napa Valley wine industry. New hotels, spas, wine-tasting rooms, and resorts cater to a recent spike in tourism.

According to the December 10, 2006, issue of the *Napa Valley Register,* city officials reported that the downtown could easily have more than one million square feet in new construction within a decade. This adds up to a more than 50 percent increase in the downtown's current nonresidential square footage. "All of a sudden . . . [there are] plenty of buyers who believe in downtown," says Bill Kampton, a commercial broker who is leasing space at Napa Square, a new office and retail complex in the downtown.

Real estate is also booming. The Napa Community Redevelopment Agency reports that real estate transactions since 1999 have totaled an extra $209 million over and above the $560 million in private investments. Craig Smith, Executive Director of the Napa Downtown Association, attributes this to the sense of security the flood-control project engendered in investors. "When the flood plan passed, that's when outside folks started buying property and investing in the future." (*San Francisco Business Times,* Vol. 22, No. 9, October 5 – 11, 2007) As of now, these investments appear to have borne fruit. Since the project's start, property values have increased by 20 percent, bucking the trend of the dot.com bust that hit Northern California. Sweetening the deal for property owners is a corresponding 20 percent decrease in flood-insurance rates.

The city's cultural and arts communities have also revived along the riverfront, supporting in addition to COPIA, the construction of the new Oxbow School of the Arts and restoration of historic buildings such as the 1914 Opera House. Equally noticeable is the number of people who have returned to the riverfront for recreational purposes, the result of cleaning up contaminated sites and installing attractive new bridges, pedestrian promenades and trails that afford good views of the river's restored nature and wildlife. Eco-tourism is also on the rise, as outdoor enthusiasts take advantage of new opportunities for fishing, kayaking, boating, and hiking along the restored river.

The result is shaping up to be the biggest transformation of downtown Napa since the city's establishment. Recent development is responsible for what the *Napa Valley Register* called in their December 10, 2006, issue, "the greatest construction surge in downtown's history." A headline in the same issue proclaimed, "Downtown Napa is getting its mojo back."

Perhaps more subtly, by mitigating the threat of recurrent, potentially devastating damage from floods, the project has finally allowed the city of Napa to benefit from the economic boom experienced by the rest of Napa Valley. This has had a profound effect on the community's sense of itself. Residents in the city no longer feel like the Napa Valley's poor, ugly stepchild. "I think downtown is now perceived as part of the Napa Valley," says Harry Price, owner of the retrofitted Napa Mill Complex and a partner in Napa Square, the new office and retail complex, as reported in the December 10, 2006, issue of the *Napa Valley Register.*

Of course, as with all economic booms, not everyone is necessarily benefiting in equal part. There is some threat that gentrification of parts of Napa is occurring due to the economic benefits of the project, displacing some lower socioeconomic groups. The community will need to keep a careful eye on the development efforts to make sure they are inclusive and sustainable.

Happily, the dialogue among diverse groups in the community that was initiated during the development of the Flood Management Plan has continued, forming a democratic and consensus-driven basis for decision making in the region on a variety of topics including housing, tourism, land use policies, and vineyard runoff, to name a few. All of these developments have resulted in a marked increase in the quality of life of Napa residents, who, just a few years ago, were demoralized by the periodic flooding and the bleak economic outlook it caused.

Perhaps the best aspect of the economic boom in Napa Valley is that — unlike in so many other places that have experienced sudden, exponential growth — it did not occur at the expense of the environment. In fact, restoring and protecting the environment is what made the economic boom possible. As Steve Kokotas, former executive director of Napa Economic Development Corporation observed:

> The most successful aspect of the project was the coming together of economic and environmental interests to satisfy the economic and environmental visions for the region. The successful marriage of the two visions can be seen strikingly on the riverfront itself. One side of the river is a beautiful, pristine natural scene. The other is an attractive, thriving urban landscape that both respects and protects the river that made it possible.

A Model Project

> This project is recognized around the world as a new way to think about flood protection.
>
> —Bill Dodd, Chair, Napa County Flood Control and
> Water Conservation District

Because of its groundbreaking approach in balancing urgent environmental, flood protection, and community needs, the Napa River project is already being used as a new model for environmental planning and disaster prevention. The project is unique in its effectiveness in building consensus among a diverse array of groups that were at complete loggerheads for decades. In bridging seemingly incommensurable philosophical differences among these groups, the project created a new, conciliatory model for planning that can be applied to any vital issue involving multiple stakeholders in conflict. The project is a model of how cities can use consensus-building to bring into balance the significant and often contradictory social, economic, environmental, and regulatory demands they face as they try to overcome persistent problems such as housing shortages, economic stagnation, environmental threats, and cultural decline.

The project also provides a model in its groundbreaking approach to balancing environmental concerns with flood management needs, proving that doing the right thing environmentally can both protect and profit the local community. Perhaps no one has benefited from seeing this approach in action more than the Army Corps of Engineers. The project resulted in nothing less than a paradigm shift in the way the Corps does business. According to Larry Dacus, a Corps member who has been closely involved in the process from the start, the project is a model for the Corps in the way it balances flood protection with environmental restoration. He also says that the agency is now using consensus-based decision making in several of its current projects (reported in the January 2005 issue of *Landscape Architecture*). Kathleen A, McGinty, Chair of the Council on Environmental Quality during the Clinton Administration, expressed a similar sentiment, calling the project a "new model of environmental decision-making."

The success of the Napa River project is being communicated to a larger audience through a variety of media. One of the best case studies of the project can be found in the book *The New Economy of Nature: The Quest to Make Conservation Profitable,* by Gretchen C. Daily and Katherine Ellison and published by Island Press. The case study includes this summary of the project:

> Napa, California, a city plagued for decades by floods, work has begun on an innovative effort to free the Napa River from its levees and dams and allow it to spill over onto its historical floodplain, proving natural flood protection. The US Army Corps of Engineers, famous for pouring concrete, began tearing it out, removing levees along a seven mile stretch. Napa residents, who have voted to raise their own taxes to pay for the plan, have seen immediate paybacks, with property values soaring in expectation of an enticing new waterfront district and a dry downtown.

Replicability

The concepts and lessons learned in the Napa project have already been transferred to consensus-based, environmental restoration projects across the United States and around the world in such places as China; Australia; Reno, Nevada; Santa Cruz, Marin, and Monterey Counties, California; and along California's American, Truckee, and San Gabriel Rivers.

Combining groundbreaking techniques in ecology, engineering, and public facilitation to successfully address the multifaceted environmental, economic, and public-safety issues of a region in crisis, the project is of wide interest to environmentalists, urban planners, public outreach facilitators, scientists, and engineers. The project's community-building and facilitation techniques serve as a model of how diverse interests ranging from public agencies, residents, business owners, and environmentalists can be brought into dialogue with each other to build a consensus and move forward on important projects benefiting everyone.

MIG, Inc., a consultant firm headquartered in Berkeley, California, was one of the key players in the process of consensus building for the Napa River project. Through projects like this, MIG has developed a tool box of methods to make the process work, including participatory charettes; one-on-one meetings; hands-on, interactive workshops; feasibility studies;

144

computer modeling; and well-designed, easy-to-understand visual materials that invite a critical discussion. These techniques each have a place and a use in the public process along with the hard science that is required to produce a viable outcome. Understanding how the social sciences, planning policy, and environmental sciences integrate with community values helps others use these concepts effectively.

Key Elements for a Successful Project

Even so, the Community Coalition model of a "Living River" planning and consensus development process does not work in every situation. At least seven key elements must be present to achieve the sort of success seen in Napa:

1. An emerging mission born from a crisis or mandate
2. Common knowledge resulting in shared meaning
3. A champion willing to take risks
4. A community of place
5. No better deals elsewhere
6. Primary parties participate in good faith
7. Multiple issues for trade-off resulting in multiple community benefits
8. Adequate resources

An Emerging Mission Born from a Crisis or Mandate

In order for the process to get started, there must be a deep and shared sense among the populace that *something must be done*. In the case of Napa, it was major floods in 1986 and 1995, combined with the unveiling of the third unacceptable Army Corps of Engineers design proposal. When natural disasters are the basis of a community crisis, you need to move quickly while the urgency is still in the minds of the locals. I suspect that when climate change begins to change weather patterns in local communities, there will be even more urgent calls for local action.

Common Knowledge Resulting in Shared Meaning

The consensus action planning process must invest in education and create a common understanding of the issues, science, and key dynamics together, so that everyone starts the process with common knowledge. This then evolves into shared meaning among the stakeholders. So often, government engineers and consultants do not adequately invest in educating the public and non-professionals about the underlying reasons behind design recommendations.

Visualization of complex principles helps in this area, as do professional facilitators. All ideas must be seriously considered, even if many of the stakeholders already know why. Everyone at the table needs to start with the same baseline information. Sometimes I refer to this success element as the need to "love *every* idea — to death."

A Local Champion Willing to Take Risks

Generally, established bureaucracies and organizations are threatened by truly open, participatory democracy planning processes. Every successful community-consensus process I have participated in has included a key elected official who leads the charge to convince the government entity and community stakeholders to take a risk in how the design and decision process needs to occur.

In the Napa experience, the "normal process" had failed three times and, given that an acceptable plan would only be implemented if two thirds of the voters agreed to a tax increase, the Flood Control District agreed to resource a community-based planning process. In doing so, it had to give up some power to the Friends of the Napa River, the Napa Valley Economic Development Corporation, and the Napa Chamber of Commerce and change the composition of its own governing body to add representation from the five cities in the county.

Additionally, there is usually at least one professional staff member who commits his or her full time and more to achieving the agreements necessary, meeting with all constituencies behind the scenes in order to identify deal-breaking positions before they come out in public planning sessions.

A planning process like this requires a few totally committed leaders who will live and breathe it for an intense and dedicated period of time.

Our experience with the flood-tax ballot for the Ross Valley Watershed in Marin County, California, shows the importance of having a local champion. County Supervisor Hall Brown (cousin of former Governor Jerry Brown) risked a lot politically in campaigning for the tax, along with a special engineer from the Flood Control District, Jack Curley, who is also a performing musician and readily accepted the job of communicating technical data in a compelling and dynamic way. The flood tax ultimately passed by 56 votes of 9,000 cast. Curley spent nights and weekends for a year staying in touch with all of the stakeholder organizations, which, if opposed to the tax in any organized way, could have scuttled the Community Coalition. You need someone who will "live" the process.

A Community of Place

It is essential that the geographic scope of the consensus planning effort is appropriate. It is not realistic to conduct a participatory process on a state or national level. Stakeholders in Florida and California are too far apart to identify with a "place" of a scale that lends itself to consensus-based planning. A watershed is an ideal geographic scope for the purpose of agreeing on a flood-protection plan. Even though the upper watershed stakeholders have different interests than the downstream residents, the Napa Valley as a whole is a "Community of Place."

Primary Parties Participate in Good Faith: No Better Deals Elsewhere

To be successful, a consensus-based community-planning process cannot allow key stakeholders to participate in an environment that allows for a better deal to be obtained elsewhere. Agreement needs to be obtained up front that participants will truly play by the rules of the process outlined by the sponsors.

A perfect example of how the lack of good faith can cripple a project may be found in the four-county Pajaro River Flood Protection Community Coalition process that has been under way since 2000. All players agreed to sit at the table to discuss how to fix their flood-control system, which is currently a system of deteriorating levees. Even with drastic river clearing, the levees can only contain ten-year flood flows. All key constituencies — the urban City of Watsonville, a strong environmental community, and strawberry and lettuce growers — sustain extreme damage in a major flood. The consensus plan is to set back the levees 100 feet on each side of the river in order to allow a reasonable low-flow channel and adequate vegetation for migrating steelhead into the upper reaches of the Pajaro River in San Benito and Santa Clara Counties.

Meanwhile, the agricultural interests believe they can get a "better deal" by forcing the boards of supervisors to set an unequivocal policy that forbids the Coalition from taking *any* land out of agricultural production. Some supervisors are willing to support that position for short-term political gain. Many growers believe that the Pajaro River is a Federal Flood Control Channel that can simply be channelized like the Los Angeles River, in spite of the Army Corps' clear statements that the federal government can only contribute funds to a project that balances flood control and environmental restoration. Both the local elected officials and the Corps are leading the major stakeholders to believe they can get a better deal outside of the Community Coalition process. As a result, there is no agreement about what to do, even after eight years of effort.

Multiple Issues for Trade-Off Resulting in Multiple Community Benefits

Consensus-driven community-planning processes tend to succeed if there is more than one issue on the table. Multi-objective flood-protection projects meet this element of success because they usually involve a multitude of issues, including flood protection, environmental restoration, transportation system improvements, land-use planning, provision for river trails and passive recreation, community health and safety, and taxation. If success *on the ground* is dependant on all of the stakeholders getting something from the process, then the more community benefits included in a project, the easier it is to achieve compromise and consensus.

Adequate Resources

Having adequate resources to tap into is crucial for a project's success. Professional planning process management, design assistance, visualization, photo simulations, adequate budget for engineering and hydrologic and hydraulic modeling (at least at the feasibility level), and community polling are all critical in achieving effective community consensus planning for large-scale projects. No one will sustain his or her participation if the process just rehashes the same limited information, meeting after meeting. A significant investment must be made by the sponsors to assure that the process will produce answers to hard questions and will not simply rely on the network of informed or uninformed opinion. It helps tremendously if there is a process to involve stakeholders in the selection of the professional support resources.

Managing large-scale consensus planning projects is never easy. If one or more of these seven elements is missing, it is difficult, if not impossible to achieve a plan that does justice to all of the issues involved or that will be supported by the majority of stakeholders. On the other hand, if, through either hard work or good fortune, all of these elements fall into place, you have a fighting chance to come out of the process with a community-supported, technically feasible, and fundable plan.

The Napa River project offers a great example of how a community of diverse and even contradictory interests can join forces and bring together all of these elements to achieve protection from nature while at the same time protecting nature. As we move into the reality of climate change and experience more and more its unpredictable, devastating effect on our communities, the need to develop successful consensus-based planning processes like the one in Napa will only become more urgent.

THE INTERNATIONAL FLOOD
MITIGATION INITIATIVE (IFMI)

Richard Gross

Dick Gross is deputy director and legal counsel for the Consensus Council (CC) in Bismarck, North Dakota, an organization that he helped to establish in 1990, and which works to develop consensus on issues of public policy in the state and region. From 1994 to 1997, Gross served as executive director of the Council of Governors' Policy Advisors (CGPA), an organization of the top four policy advisers appointed by each governor. From 1972 to 1993, he served in North Dakota state government in numerous positions, including attorney for the ND legislature, assistant attorney general and, for his last eight years in state government, as legal counsel and policy director to the governor. He has served as chair of the Staff Advisory Committee of the Western Governors' Association (WGA), and chair of the Energy and Environment and the Canada–US Free Trade Agreement Staff Advisory Committees of the National Governors' Association (NGA), as a member of the Directorate of the American Planning Association (APA), and the Public Policy Committee of United Way of America. He is a graduate of Marquette University and the University of North Dakota School of Law.

Introduction

The Red River flows north, originating in northwest South Dakota and flowing into Lake Winnipeg, about 600 river miles away. While it may not have required an ark, one would have come in handy during the spring of 1997 in eastern North Dakota and western Minnesota. That was when the Red River flooded in epic proportions, submerging the cities of East Grand Forks, MN, and Grand Forks, ND. Approximately 60,000 people were driven out of their homes in the two cities. Both cities were virtually wiped out.

The building containing the offices of the *Grand Forks Herald*, the newspaper for both cities, burned to the ground as waters swirled around it. They moved their operations to a small community out of the flood zone and continued publishing the newspaper, using printing presses in Minneapolis. They never ceased publishing or delivery, and their coverage of the disaster led to a Pulitzer Prize award for the editor and publisher.

FIGURE 5.2 Grand Forks, ND, April 1, 1997 — Aerial of a flooded Grand Forks neighborhood after the Red River floodwaters came though Grand Forks. FEMA/ Mike Rieger

That effort was emblematic of what were heroic efforts to save lives and restore the lost property afterward.

Not a life was lost to the flood. And now, ten years later, the cities are fully rebuilt with flood protection levies on both sides of the river.

But that is not the story this essay will cover. What happened in between to help restore these cities and provide greater protection from and mitigation of future flooding on the "Red" was a collaborative effort involving two countries, two states, a Canadian province, and the respective federal governments of the United States and Canada.

Initial Efforts

Within weeks after the flood and the accomplishment of rescue efforts, as Grand Forks and East Grand Forks began their long rebuilding, a small nonprofit in Bismarck held its bimonthly board meeting. Board members, representing major organizations in North Dakota, had all seen the devastated communities, most in person. They persuaded the staff of the Consensus Council, the small nonprofit, that those cities needed help in community rebuilding.

Partnering with the *Grand Forks Herald*, the Consensus Council held community meetings in churches, schools, and community centers around

FIGURE 5.3 East Grand Forks, MN, April 1997 — Welcome to East Grand Forks, as a Coast Guard crew patrols the flooded neighborhoods of the city. Photo by David Saville/FEMA

the area and received significant input about what the citizens wanted their communities to look like as they were rebuilt. The citizens noted, for example, that there was not sufficient affordable housing in the two communities before the flood and that building more such housing should be a focus of the reconstruction efforts. They also noted that large shopping centers had diverted the citizens from their downtown areas, and the downtowns should be rebuilt in a way that re-attracted citizens there as the communities' gathering places. Although the cities' fathers (and mothers) would probably not give a lot of credit to those initial meetings, the council's work was recognized and its recommendations, as received from the citizens, afforded credibility. That was the beginning.

What the council staff learned in those initial efforts was the degree of blame and mistrust of their own and other officials at the local, state, provincial, regional, and national levels in the United States and Canada . . . and those officials' mistrust of each other. Following the flood, blaming was endless. The Corps of Engineers messing around (or not) with the river over decades had set it up for disaster. The two states which the river borders, North Dakota and Minnesota, had built levees and dikes that "pushed" the river either too far east or too far west, or forced it into too narrow a channel so that, when the levees and dikes ended, the river waters gushed far overland, consuming farms and smaller communities along the way and, eventually overcoming dikes, levees, and sandbags that had been built to protect the cross-border cities of East and Grand Forks.

FIGURE 5.4   Grand Forks, ND, May 1997 — Aerial view of a flooded property with a house and its immediate lawn like an island in Grand Forks. FEMA/ Michael Rieger

But there were many others blamed for the catastrophe. The national, regional, and local weather services gave inaccurate reports about the river's crest. Farmers had destroyed too many wetlands; so there was little to hold the floodwaters back on the land or tributaries that feed into the Red. Tributaries, too, had dikes and levees. And the province of Manitoba had created a road (North Dakotans called it a dam) perpendicular to the flow of the river, just north of the U.S.–Canada border with culverts too small to let enough water through. As a result, waters backed up for miles, slowing the river's flow north.

The Consensus Council sought and received small grants from FEMA and the German-Marshall Fund, totaling $23,000 and, with those funds, through work with the embassy of the Netherlands in Washington, D.C., developed an itinerary for a week in the Netherlands, especially for water officials at those various governmental levels in both countries. The Council's feeling was that, if any country knew about flood control and mitigation, it had to be the Netherlands, 40 percent of which has been recovered from the ocean, and which works continuously to keep the ocean at bay.

The Council invited 30 critically positioned water officials at the local, state, provincial, regional, and federal levels to join it on a bus tour through the Netherlands to visit small, medium, and large flood control/mitigation efforts with officials of the Rhine Commission, which works internationally

FIGURE 5.5 East Grand Forks, MN, April 1997 — Dave Pauli of the Humane Society of the United States proudly shows off a rescued pet. Animal-rescue operations in the Grand Forks area continued for several days. Photo by: David Saville/FEMA

to control the waters of the Rhine as it meanders through several European countries. Most of the officials from the United States and Canada who joined the bus tour obtained agency funding to join the trip. Some needed additional funding, and the small amount of grant funding was used to help them. Council staff took these officials on a tour throughout the Netherlands on a bus for a week and facilitated their meetings.

People who had been blaming each other for the flood ate together, traveled together, worked and learned together for that week and many became — and remain — friends. When the group returned from the trip, in May of 1998, those 30 participants presented at a "summit meeting" in Grand Forks. The trip's participants were panelists who presented information to 90 other officials gathered there about what they had learned on their tour of the Netherlands. It was a historic gathering, which many left with a new vision of what could be done for the future of the Red River Valley.

Establishing the International Flood Mitigation Initiative (IFMI)

Two regional FEMA officials who accompanied the other participants on the trip through the Netherlands wrote a letter to the national FEMA office that was highly complimentary of the trip and the facilitation efforts that had been done during the trip. They wrote that this effort should not stop with

FIGURE 5.6 Grand Forks, ND, May 1997 — Aerial view of Grand Forks neighborhood and the flooded Red River of the North. The levee in the foreground was topped with sandbags but breached, flooding the area near the river. FEMA/ Michael Rieger

the trip. The national FEMA office invited a proposal from the Consensus Council to develop and facilitate a multijurisdictional drought-mitigation planning effort involving senior policy makers from both countries, at all levels of government. FEMA awarded a grant from its "Project Impact" funding of $235,000, and the province of Manitoba awarded an additional $100,000 to carry out the effort over what the council believed would take eighteen months (the project would be extended to two years so that consensus on all major recommendations could be achieved).

The Consensus Council invited facilitators from Manitoba to join its team so that participants from both countries would feel that they were equally represented by staff and facilitators who understood the somewhat different cultures of the two countries.

So, over two years, beginning in the fall of 1998 and ending in November of 2000, a very difficult process, the outcome of which would be broad consensus on many significant issues, was carried out. Because I was the lead facilitator for the process, I was asked to write this essay. Other facilitators and staff contributed greatly to the effort, and without their expertise the process would never have been as successful and may, in fact, have withered many different times along the way. Their moral support alone — for me and for the participants — was essential to bring the process to a successful conclusion. Probably the most important lesson we learned, aside from the critical importance of rebuilding personal relationships at the outset with the bus ride through the Netherlands, was how important the teamwork and hard work by the staff was to a process of this magnitude.

Different laws between the states and the states and province were problems, as were different levels of responsibility and agencies in the two countries. It was necessary for the participants to learn that there were simply not parallel agencies in the two countries that could interact to develop and implement agreements. There was no Canadian counterpart to FEMA, for example, which was obviously integral to the rebuilding efforts on the U.S. side of the border.

Creating a Core Group

We began the agreement-building process by inviting a group that we believed would be core, not only to developing but also to implementing any agreements reached. We also wanted such a core group that would attract other essential "stakeholders" to the process. And it worked like a charm. The Regional Director of the Army Corps of Engineers; two former North Dakota governors, one whom was then head of the Independent Community Bankers Association in Minnesota, the other a retired North Dakota governor living in Fargo, ND; the majority leader of the Minnesota Senate, the Manitoba clerk of the Executive Council, Office of the Premier, and Manitoba's deputy minister of conservation were all invited first. They all agreed to participate. Had they known how long and difficult the process would be, they might have declined.

But the importance of the core group was another lesson learned early. With people of that stature willing to participate in what we knew would be at least a year-long process, it was relatively easy to bring others to the table — core FEMA officials at the regional and national levels, a foundation president, the North Dakota Insurance Commissioner, a regional EPA Assistant Administrator, mayors, environmentalists, conservationists, legislators, college officials, and other significant water experts joined on until we had 30 participants. They gathered for the first time in November of 1998 and created the International Flood Mitigation Initiative (IFMI).

Process

During the 14 meetings held, eventually over 24 months (approximately one every eight weeks), enthusiasm and confidence that there would be

FIGURE 5.7 Grand Forks, ND, April 1, 1997 — A flooded neighborhood with cars still floating due the floodwaters from the Red River of the North. FEMA/ Michael Rieger

agreements waxed and waned. At times the effort floundered, but it never stopped. When one staff person or several participants were prepared to throw in the towel, other staff and participants came to the rescue. When no one was certain of the direction, community meetings and public support for the effort rejuvenated the process. In the end, everyone agreed that the issues were too important and that not gaining broad consensus would be the worst of all possible outcomes. So the work continued.

Each meeting began with an educational session in order to bring all participants to a similar level of general expertise on river history, ecology, and hydrological aspects; on weather forecasting; on the political, legal, and public policy issues each participant faced. Experts from throughout the valley, at the regional, national, and even international levels presented at these meetings. In retrospect, it was an extraordinary effort representing incredible commitment by participants and staff.

The Initiative concluded with a meeting at which the governors of North Dakota and Minnesota, a top representative of the governor of South Dakota and the premier of Manitoba met for the first time in history on water issues with the IFMI members, heard their report and recommendations, and agreed to a Memorandum of Understanding (MOU) about the Red River that remains in place today. It was obviously a historic meeting. But what happened between the formation and the last meeting with the governors and premier in Fargo, ND, is a significant part of the rest of the story.

Shared Understandings

Integral to the agreements developed on vision, goals, objectives, strategies, policies, projects, and partnerships was an early development of shared understandings:

- **One watershed community.** The Red River Basin is one community that transcends political boundaries, and the risks and benefits of the Red River and its tributaries are shared by Manitoba, Minnesota, North Dakota, and South Dakota.
- **New partnerships.** New partnerships among the government, the private sector and nongovernmental organizations (NGOs) are essential to reduce flood damages.
- **Basin-wide participation, coordination, and cooperation.** A participatory basin-wide mechanism is needed to coordinate cooperative flood-mitigation efforts among all political jurisdictions, affected constituencies, and the public.
- **Importance of pre-disaster mitigation.** The Red River and its tributaries have and will continue to experience flooding of 1997 levels or greater in the future. Future damage to people, property, and environment and costs to taxpayers can be reduced by investing in effective mitigation measures before the next major flood event occurs.
- **Flood-resilient communities and region.** Achieving flood resilience will require increased public awareness and understanding and innovation in public policy, institutions, physical structures, landscape management, and land use.
- **Accomplishing multiple objectives.** Flood mitigation can and should be linked to environmental enhancement, economic development, and community well-being. The Red River and its tributaries present an untapped resource for flood mitigation, conservation, and local and regional economic development through recreation and tourism.
- **Implementation.** Effective implementation of flood mitigation requires funding, technical resources, and cooperative institutional frameworks.

FIGURE 5.8 East Grand Forks, MN, April 1997 — As the water receded, traffic signs began to appear near the Kennedy Bridge in East Grand Forks, MN. Photo by FEMA/ David Saville

On the foundation of these shared understandings, the agreements to take action were based.

The first word that presented a problem was "mitigation." What does it mean? After extended discussions, the participants agreed that, at least in terms of flooding on the Red River, the word meant, "systemic and sustained actions that substantially reduce risk of harm to human life, property, and the environment." In other words, as indicated in the "shared understandings," the participants understood that there will always be another flood on the Red River and/or its tributaries and that what was necessary was to take sustained actions that would substantially reduce damages and losses from such flooding in the future.

Vision and Mission Statements

Based on their shared understandings and definition, the group developed vision and mission statements.

The vision of the IFMI participants was: "By the Year 2010, the community of the Red River Basin has addressed flooding through mitigation that achieves significant flood damage reductions goals while enhancing economic, social, and ecological opportunities."

FIGURE 5.9 East Grand Forks, MN, April 8, 1997 — A man walks along the flooded roadbed of Highway 2 toward the Kennedy Bridge. Photo by Dave Saville/FEMA News Photo

FIGURE 5.10 East Grand Forks, MN, April 18, 1997 — Residents of East Grand Forks are helped into National Guard trucks as the city is evacuated. FEMA/ David Saville

The mission of IFMI was, "To promote and develop achievable and action-oriented flood mitigation goals and implementation strategies by engaging citizens, their communities, and governments."

Goals

The two major goals for which objectives and strategies were developed were:

- To develop and implement a basin-wide framework for flood mitigation
- To protect people and property while enhancing the environment, economy, and communities.

Major Agreements

Based on these vision, mission, and goal statements, five objectives were developed that led to the development of 14 major agreements for future action. The 14 agreements, in summary form, were:

- To develop a Memorandum of Understanding (MOU) among the government leaders and a revised Red River Basin Commission
- To convene an ongoing legislator dialogue
- To establish a broadcast media partnership
- To establish a print journalism partnership
- To create a Red River Basin Institute to do research, mapping, and education
- To establish a center for watershed education
- To conduct a reconnaissance study of basin-wide flood-mitigation options
- To provide technical assistance to communities in the basin for flood mitigation
- To establish a "Greenway on the Red"
- To establish an institute for floodplain architecture
- To increase lender compliance with U.S. flood insurance laws and regulations
- To recommend changes to the U.S. system of flood insurance
- To secure passage of the "Farmland Stewardship Initiative" in the U.S.
- To establish a multistate conservation reserve enhancement program

Implementation Strategies

Many specific strategies were adopted to implement these agreements; and participants agreed to take them on or hand them over to more appropriate persons, agencies, or organizations. While all of these agreements have been implemented, some to a greater extent than others, the remainder of this essay will focus on the implementation actions with which I am most familiar.

Establish a Legislators Forum — Since November of 2000, the time of the completion of the IFMI consensus building effort and final report to the governors and premier, a steering committee, the Legislators Forum, comprised of two legislators from each of the four jurisdictions, has been in place through which annual meetings of thirty-two legislators have taken place. The Consensus Council has staffed and facilitated the steering committee and annual Legislators Forum meetings. The Forum has grown in its consideration of issues common to the four jurisdictions to move well beyond water-related issues. However, each of the seven annual meetings held since 2001 has had as one area of focus water, climate, and, recently, drought mitigation.

In addition, over the seven-year period, the "delegates" from each jurisdiction — eight pan-partisan legislators from each — have considered and developed general agreements on a wide variety of issues, several of which have had significant results. (See Sidebar 1.)

Many other issues have been considered during the seven-year life span of the Forum, and I believe the relationships and agreements that have resulted have had a significant impact on legislation and working together between the four jurisdictions in the region. What is perhaps most significant is that these meetings and this agreement building have continued in spite of significant executive branch issues that have gone on relative to a continuously rising Devils Lake in North Dakota and the building of an outlet from Devils Lake, which moves filtered water into the Cheyenne River, a tributary of the Red River. While the Canadian and Manitoba governments and the state of Minnesota have strongly opposed the outlet and drainage into the Red, and countless headlines have documented the depth of those disagreements, the Legislators Forum has remained a place, where legislators from these three jurisdictions and South Dakota can come together to discuss issues and develop agreements. To date, the delegates have chosen to steer clear of the Devils Lake issues.

The governors of the three jurisdictions and the premier have met three times since the agreements were developed relative to Red River

---

Sidebar I
General Agreements Reached by the Legislators Forum

Tourism — a two-Nation Tour program has been developed including all four jurisdictions and is flourishing.

Renewable and Clean Coal Technologies — a Powering the Plains consensus-based effort has developed an "Energy Roadmap" and developed agreements among a very diverse group for the region to transition from primary reliance on traditional coal to a wide variety of renewable energy, including hydrogen, and clean coal technologies, including sequestration of CO_2, which is already being done by the Dakota Gasification Company in North Dakota, which sends CO_2 via pipeline to Saskatchewan oil fields to provide for enhanced oil recovery.

The Western Hemisphere Travel Initiative (WHTI) — the delegates have agreed and contacted their respective federal governments relative to their agreement that the current passport requirement for those who travel into and from Canada to the United States should be eliminated.

Water — the Red River Basin Commission (noted in the first agreement above) has replaced and embraced disparate groups that were trying to make decisions relative to management of the Red River, and, based at least in part on agreements by these legislators, is in the process of developing agreements on water management and sharing, particularly in the event of a drought.

Broadband Technology — the delegates agreed in their recent meeting in Pierre, SD (July of 2007), to support "universal service" so that those located in rural areas will have as much access to broadband service as research universities at reasonable prices.

Alternatives to Incarceration — in a movement away from the continuous building of new places to incarcerate a continuously growing number of inmates, the Forum has discussed and agreed upon development of treatment centers, education, and prevention efforts, drug courts, and other forms of diversion from the traditional criminal justice system.

Valley flood and drought mitigation. The MOU they agreed to was not nearly as extensive as the one developed by the IFMI participants; but the fact that they have met has helped to keep Red River and tributary issues on the table and helped to implement other agreements, such as those referred to below:

Establishing broadcast media and print journalism partnerships on water issues — In the aftermath of the 1997 Flood, the International Flood Mitigation Initiative (IFMI) established its goal to *provide comprehensive, continuous, consistent, complete, current, and accurate flood forecasting, flood mitigation, flood preparedness, flood response, and when necessary, flood recovery information to the people of the Red River Basin.* To fulfill this goal, Prairie Public Broadcasting has established and maintained **RiverWatch** — using cross-border and cross-media partnerships — to provide flood preparation, recovery, and mitigation news, as well as news about other drought and water quality issues facing the Red River.

These informational resources are provided through television, radio, Internet, and public outreach. **RiverWatch** is viable and respected because, together with its research, content, and relationships, Prairie Public Broadcasting has also contributed its technical capacity to gather and distribute information along the entire length and breadth of the Red River Valley. (See Sidebar 2.)

Sidebar 2
RiverWatch Project Elements

Over the past five years, RiverWatch elements have included:

1. **Television:** Weeknight news updates air every night on Prairie Public Television during peak flood months (March–May), with features on flood preparation, recovery, and mitigation efforts by individuals and communities. In addition, year-round television segments cover a wide variety of relevant topics, such as ice rescue diving, emergency levee construction, fishing, invasive aquatic plants, and chemical contaminants in the Red River.
2. **Radio:** News reports and features are integrated into *Morning Edition* and *All Things Considered* on North Dakota Public Radio. In addition, *Hear It Now,* North Dakota Public Radio's daily public affairs show, airs interview and discussion segments together with **RiverWatch** features.
3. **Web:** Year-round river- and flood-related news updates are posted daily to the **RiverWatch** Web site at www.river-watchonline.org. In addition, **RiverWatch online** provides comprehensive links to authoritative flood/water management and educational resources, together with flood preparation and recovery resources, information, and links. Updates

163

and information on the Greenway on the Red and other educational opportunities along the Red River are also posted on **RiverWatchonline.org.** During times of flooding, the people of the region consistently visit www.riverwatchonline.org to get the latest weather, river levels, news, and flood recovery information and resources, with an average annual visitation of over 350,000. The **RiverWatch online** site features more than one thousand pages of public information.

RiverWatch continues to gather and share information, news, and resources with more than forty partnerships of Red River Valley media outlets, communities, municipalities, and federal, state, and provincial agencies. The **RiverWatch** team continues to pursue additional partnerships with agencies, organizations, and municipalities in the United States and Canada.

In addition, a documentary entitled *Red River Divide* has been produced and disseminated throughout the Basin. *Red River Divide* investigates the Red River Valley's geology and landscape, the history of flooding and flood mitigation and the tremendous benefits the Red River provides. The documentary also explores what preparations are under way for drought relief and protection of the river's water quality. This 60-minute documentary will be rebroadcast multiple times on Prairie Public Television, with an audience across North Dakota, northwestern Minnesota, Manitoba, and eastern Montana.

Through these collective efforts, **RiverWatch** will continue to provide a significant service to the thousands of resident living in the Red River Valley — before, during and after flooding events.

Establish a **Greenway on the Red**: One of these critical initiatives has been the establishment of a continuous Greenway along the Red River and its tributaries that links the people and communities of the Red River Basin from Lake Traverse in South Dakota to Lake Winnipeg in Manitoba. (http://www.prairiepublic.org/features/riverwatch/greenway/greenway.html)

The **Greenway on the Red** was formed as a nonprofit organization to coordinate the development of a 600-mile/960-kilometer continuous Greenway system on the Red River of the North and its tributaries. The Greenway strives to mitigate flood damage and protect people through education and partnerships that enhance the economy, environment, and

communities of the Red River Basin. In addition to mitigating the destruction and hardship caused by inevitable flooding in the Red River Basin, the
Greenway is providing multiple on-the-ground benefits through riparian
restoration; water quality enhancement; farmer/landowner incentives;
community strengthening; and increased recreation, tourism, and economic development. (See Sidebar 4.)

The **Greenway on the Red** has been formulated to work closely with
the other strategies and projects of the International Flood Mitigation
Initiative. Greenway activities also have and will continue to involve proactive and strategic partnerships with multiple entities in the Red River
Basin to enhance Greenway development. In addition to ongoing linkages in North Dakota, South Dakota, Minnesota, and Manitoba, evolving
partnerships include:

- Red River Riparian Project
- Red River Basin Commission
- Red River Management Consortium
- ND Game and Fish Department
- ND Department of Parks & Tourism
- State and National Park Services
- Conservation Fund
- Historical Society (MN and ND)
- City and County Commissions
- Resource Conservation & Development
- Federal Agencies: Fish and Wildlife Service, Bureau of Reclamation, Natural Resources Conservation Service, and others
- Private Landowners

Create a Red River Basin Institute to do research, mapping, and education and establish a Center for Watershed Education — International Water
Institute: The International Water Institute (www.internationalwaterinstit
ute,org) was established as a result of the International Flood Mitigation
Initiative, to provide a forum for research, public education, training, and
information dissemination relating to flood damage reduction and water
resource protection and enhancement in the Red River Basin. The Institute
delivers applied research, education, and outreach programs through the
Institute's Center for Flood Damage and Natural Resources and Red River
Center for Watershed Education. (See Sidebar 3.)

Sidebar 3
International Water Institute Programs

The International Water Institute's science-based Center for Flood Damage and Natural Resources provides a biennial water conference that convenes numerous water resource experts, as a mechanism to keep the Red River community apprised of relevant and current research in the basin. The center also partners with North Dakota State University to host the Red River Basin Decision Information Network (RRBDIN — www.rrbdin.org). RRBDIN was originally developed during the 1997 flood by the International Joint Commission, the US Army Corps of Engineers (USACOE), and the Federal Geographic Data Committee to serve as a one-stop portal for water resources management information. It continues to be a vital and interactive tool for the use and exchange of scientific information and research.

The Institute Center places an important emphasis on applied research in the Red River basin, as exemplified by a recently completed nutrient and ion water quality study of the Red River for the International Joint Commission (IJC). The Institute worked with IJC senior science advisers and faculty at Concordia College to investigate nutrient and ion data dynamics in the Red River of the North. The study will be used by the International Red River Board to addresses Lake Winnipeg concerns and develop nutrient objectives at the International Border.

The International Water Institute was "born" from the need for accurate information and research relevant to flooding. The Flood Forecast Display Tool (FFDT) exemplifies this objective. Operational in the spring of 2006, the FFDT generated flood forecasts well into July, and operated flawlessly. Over 2,000 unique hits per month were registered when flooding was occurring. The FFDT is a pilot project that was undertaken to address many outstanding technical issues related to generating interactive maps through the Internet using near real-time flood forecast information (www.rrbdin.org) from the National Weather Service. The FFDT is the first of its kind and a template that will eventually be delivered across the entire Red River Basin.

The critical need for uniform mapping across the Red River basin was clearly elucidated during the IFMI process. The Institute's Red River Basin Mapping Initiative (RRBMI) has been designed to meet that need. It is expected to last three years, with data collection beginning in spring 2008. The $5 million RRBMI project will collect high-resolution elevation data across the entire U.S. portion of the Red

River Basin (39,400 square miles). The RRBMI will be one of the largest high-resolution elevation data collection projects undertaken in the U.S. When completed, high-resolution digital elevation data will be publicly available through the Internet. Benefits from this project will be widespread, with positive impacts in nearly every economic sector and level of government. Most important, the Institute and partners believe the information will profoundly impact how water and resources management decisions are made and defended by society.

As an outcome of the EPA Region 8 Bio-assessment Workgroup, the Institute has developed a Web site to serve as a one-stop information source and exchange for Environmental Protection Agency (EPA) Region 8 Wetland Workgroup members, EPA staff, academia, consultants, and others seeking information about wetland bio-assessment efforts. The Institute will host a Region 8 EPA Wetland Monitoring and Bio-assessment Workshop in Rapid City, SD, in April of 2008.

The Red River Center for Watershed Education has evolved into an innovative educational force within the Red River basin, providing students, teachers, and citizens with a broad range of opportunities to learn, while contributing to the overall knowledge and understanding of the Red River. Through the Basin Volunteer Monitoring Network, which has been built upon Minnesota's RiverWatch program, the Institute is working with partners in MN, ND, and MB to establish a basin-wide network of volunteer monitors to collect and distribute water-quality data from rivers and streams in the Red River Basin. These data are collected using common collection methods that are scientifically defendable. An innovative and interactive centralized database has been developed (http://riverwatch.umn.edu/). In addition, the Center has implemented the River of Dreams, which uses launched small cedar canoes to educate participants about the Red River watershed and instill a sense of place and community (http://www.internationalwaterinstitute.org/dreams/index.htm).

Through the Understanding the Science Connected to Technology (USCT) grant from the National Science Foundation, the Center has had a profound impact on watershed science capacity and interest of the students and teachers involved in the program.

The Institute provides watershed-relevant continuing education opportunities through the annual Red River and You Teachers Institute. Each year, teachers from around the Red River Basin spend a week at relevant sites in the field, learning about Watershed Education. Most of the participants received annual graduate credit.

Sidebar 4
Greenway on the Red Implementation Tasks

The establishment of a contiguous Greenway has been focused on four major implementation tasks:

- **Develop a centralized inventory and assessment of individual, ongoing urban and rural Greenway initiatives as a base for the provision of support and linkage services.** Efforts to compile and delineate existing Greenway resources were preceded by strategic planning to define what specific elements comprised the Greenway. Specific planning and design elements incorporate flood damage reduction, water quality enhancement, channel/bank stability (to control erosion and slumping), contiguous habitat and biodiversity corridors, economic development, and cultural enhancement and preservation.

- **Develop a restoration and private landowner outreach strategy for on-the-ground Greenway. Greenway on the Red** has compiled a Riparian Handbook that summarizes best management practices, strategies, and funding sources for riparian restoration and conservation appropriate for Greenway implementation. In addition, over the course of Greenway strategic planning and implementation, it was recognized as essential to have tangible, ongoing, site-specific restoration "examples," to demonstrate and actualize what the Greenway should be. To that end, **Greenway on the Red** partnered with the Red River Riparian Project to physically design and restore multiple acres of riparian habitat along the Red River of the North and its tributaries. State-specific initiatives have also been vitally important to Greenway implementation. Following the recommendation of the governor of North Dakota, the North Dakota Game and Fish Department announced the acquisition of 6,000 riparian acres along the Red River and its tributaries for the Greenway in North Dakota. In Minnesota, the state Board of Soil and Water Resources successfully initiated a Conservation Reserve Enhancement Program (CREP), which funds riparian restoration along the Red River and its tributaries, up to a 25,000-acre limit.

- **Increase awareness, participation, and educational opportunities in the Greenway**, in conjunction with the Broadcast

168

Media Partnership and the Red River Institute for Watershed Education. The **Greenway on the Red** has benefited from a long-standing relationship with Prairie Public Broadcasting's (PPB) Broadcast Media Partnership and the Red River Institute for Watershed Education. Ongoing Greenway information dissemination has been effectively achieved through the Broadcast Media Partnership's "RiverWatch" website (http://www.riverwatchonline.org/greenway/index.html), which also provides a broad range of information related to the Red River, its history, its natural resource values, and its propensity to flood. In addition, the **Greenway on the Red** has benefited from numerous outreach articles, presentations, brochures and other means of informational outreach. Greenway involvement in county-based planning initiatives has focused on strategies that maximize concurrent benefits for bird/wildlife habitat, riparian/wetland restoration, flood damage reduction, and water quality. Assistance has also included recommendations for the establishment of riparian development setback requirements and minimum vegetative cover, including planting recommendations.

- Research linkages necessary to delineate and monitor the evolving Greenway through the International Water Institute, through its Centers for Flood Damage Reduction and Watershed Education. It is imperative that the work carried out on the ground through the **Greenway on the Red** be credible and science-based. To that end, the International Water Institute, also established as an IFMI initiative, has functioned as the Science Technical Committee for the Greenway since its inception. Many of the science-based activities of the International Water Institute have direct benefits to Greenway development and implementation. The Institute's Red River Basin Mapping Initiative (RRBMI), which will collect high-resolution LIDAR (Light Detection and Ranging) elevation data across the entire U.S. portion of the Red River Basin (39,400 square miles), will be vital in the prioritization and subsequent restoration of Greenway sites, to achieve flood resiliency and as well as contiguous wildlife habitat corridors. The partnership between the International Water Institute and the Greenway on the Red has been proactive and synergistic.

IFMI as a Model

In its concluding statement, "IFMI as a Model," the participants stated: "IFMI's collaborative, participatory approach to developing basin-wide goals, objectives, targets and measures to jointly designing and implementing model organizations, institutional partnerships, policies and projects can readily be applied to other contexts." FEMA and IFMI staff have received inquiries about IFMI from other regions of North America, and FEMA was invited to give an in-depth presentation about the IFMI process and its results to an international hazard-mitigation conference held in Cairo, Egypt. Its model has also been used in post-Hurricane Katrina work.

The following elements have proven essential to the creativity and productivity of the IFMI process:

- Affected parties participate as equal partners, including federal, provincial, state, and local governments; business and industry associations; local water and agricultural interests; higher education; conservation and environmental and disaster relief organizations; and charitable foundations.
- Participants solicit and respond to citizen and constituency input.
- Consultation is transboundary in scope, treating all jurisdictions fairly and equally.
- Public, private, and nongovernmental partnerships are built through and during the process.
- Decisions are made by consensus.
- Innovative and imaginative solutions are sought, in which all parties can win — before resorting to compromise, in which some or all parties may lose.
- Objectives include measurable targets and timeframes for implementation.
- Decisions are linked to practical implementation to achieve desired results.
- Mechanisms for implementation are developed and institutionalized during the process.
- Stakeholders participate directly in the implementation of new initiatives.
- Financial and other resources are identified and developed during the process to enable and sustain implementation.

IFMI has sought to build an enduring legacy of practical initiatives that will contribute to flood mitigation in the Basin long after IFMI itself has

FIGURE 5.11   East Grand Forks, MN, April 1997—Olson Drug in downtown East Grand Forks, MN. Photo by FEMA/David Saville.

come to an end. The IFMI process benefited greatly from the opportunity to learn about successful models of river basin water management and flood mitigation in other regions of North America and internationally. Just as IFMI participants took advantage of others' experience, they believe that their work can also serve as a model for other organizations and institutions in the Red River Basin, elsewhere in North America and even overseas.

A recent news article in the *Grand Forks Herald* announced, "Grand Forks property owners stuck with high flood insurance premiums will be able to breathe a sigh of relief a week from today, city officials said Wednesday, Aug. 2 is the day the Federal Emergency Management Agency officially recognizes the city's dikes are done and can protect homes and businesses from the next 100-year flood. Mayor Mike Brown said the fact that this is happening in 2007 is remarkable."

While none of the IFMI participants would claim credit for the fact that the dikes are completed and able to protect against a 100-year flood, the story is an example of the kind of work that has been done in the Red River Valley Basin since the 1997 flood, through collaborative efforts among units of government, the private sector, and nongovernmental organizations. Ultimately, that is what the IFMI participants did and agreed was necessary throughout the Basin. Through the agreements reached in the IFMI process and the example it provided, much has been accomplished to mitigate future flooding, and much more will be accomplished to address both flood and drought issues in that Basin and elsewhere in the four jurisdictions.

SEATTLE PROJECT IMPACT:
STRENGTHENING A REGION'S CAPACITY TO
MITIGATE, RESPOND, AND RECOVER

Inés Pearce

Inés Pearce launched Pearce Global Partners (PGP) in 2006 to address the needs of government, business, and communities to help reduce the potential for devastating loss of life and property resulting from natural and human-caused disasters. As an expert in public-private partnerships, in 2007 Ms. Pearce represented the World Economic Forum at the United Nation's Global Platform for Disaster Risk Reduction in Geneva, Switzerland. In response to the 2007 California wildfires, Ms. Pearce was selected by the U.S. Chamber of Commerce Business Civic Leadership Center to be a liaison during times of disaster and facilitate long-term recovery efforts. Prior to PGP, Inés was appointed as Seattle Project Impact Director for the City of Seattle Emergency Management in December 1998, managing four mitigation programs that provided resources for safer schools, homes, and businesses, and better hazard maps. Ms. Pearce is current president of the Contingency Planning and Recovery Management (CPARM) group, president of the nonprofit Disaster Resistant Business (DRB) Toolkit Workgroup, and on the board of directors for the Cascadia Region Earthquake Workgroup (CREW). The author would like to thank Ryan Walker for his support and insight in the creation of this chapter.

In emergency management, we are always racing against the "ticking clock" of the next disaster. As a means to tackle this in 1997, Project Impact was presented by the Federal Emergency Management Agency (FEMA) as a public-private partnership initiative with the purpose of fashioning disaster-resistant communities. Asked to be one of the seven pilot communities, the city of Seattle looked upon this initiative as a rare opportunity to bind together businesses and community organizations that had not yet had an opportunity to collaborate, as well as to significantly focus on disaster-mitigation issues in order to reduce potential loss of life and property.

Seattle had been selected as a pilot community due to the reputation of the then-existing government mitigation efforts made by the city, especially the staff of the City of Seattle Emergency Management and their innovative approach toward the involvement of the private sector in their community. Just as in every jurisdiction around the United States, in Seattle the attention to mitigation efforts was limited to the times immediately following a disaster and then only focused on government buildings or structures. The government-specific mitigation approach up to this point was a more post-recovery approach than that of true risk reduction.

In FEMA's new Project Impact lay the chance to engage not only the public sector, but the entire community to determine first what vulnerabilities were most important, and then which local solutions were to be applied with regional, state, and federal enhancements and support.

FEMA's approach with Project Impact was unique for a federal agency. They came to local government with money and broad guidance but no mandates on exactly how the funds were to be spent. Traditionally, government grants were incredibly cryptic and difficult to navigate. While initially reluctant of potential "strings" tied to the funds based on previous experiences with federal programs, the City of Seattle Office of Emergency Management found the initiative compelling and agreed to participate as a pilot community. As the FEMA grant funds were intended to be used as "seed money," Seattle Project Impact began working in earnest in November of 1997 with every intention to stretch each penny as far as possible and to leverage all partners in order to institutionalize mitigation, rather than simply spending the all of the seed money.

The goal was to build a disaster-resistant city under the leadership of the Seattle Office of Emergency Management (OEM), but OEM and its partners quickly recognized that actual disasters do not follow jurisdictional boundaries, especially when considering the greatest threats to that region; earthquakes and landslides. So, from the beginning, partners were gathered from around the region to participate in program development that could be broadly expanded. The seed money was intended to help kick-start an effort. Those of us in Seattle OEM realized that in order to truly perform risk reduction and make lives safer, our solutions had to come in the form of sustainable, long-term programs that involved the entire local community.

Seattle did this through the use of three guiding principles announced at every meeting in order to set the tone. Seattle Project Impact programs had to:

1. Be of substantial benefit to the community
2. Meet partners' timelines
3. Be easily exported to other jurisdictions

These guiding principles became the foundation of all programs developed.

To facilitate the development, OEM hired a full-time director to lead the overall efforts, partners, committees, grant management, and momentum with the assistance of Seattle Project Impact partners, committee chairs, and other OEM staff. While initially designed as a Project Impact grant-funded position, the city used other grant funds to cover the position and freed the use of the monies for other activities. Seattle's experience and perspective shared with other Project Impact communities explained that in order to be successful, a community must hire at least one full-time person to coordinate activities and partners. By 2000, there were over 250 Project Impact communities nationwide, some of whom were fortunate to hire a staff, but, because mitigation was still new on the radar of many, the vast majority of communities only had one person working part-time. In Seattle, having a dedicated person allowed OEM to elevate mitigation so that each phase of emergency management was covered by at least one professional staff member and demonstrated the need to continue Project Impact activities long after grant funds were expended.

DEVELOPING THE PARTNERSHIP AND PROGRAMS

The first step undertaken was the task of defining the partnership. While Project Impact was a public-private partnership initiative, in such partnerships it is critical not to limit the partnership to only the governmental and business sectors. This is intended to be a community-wide partnership with involvement at every level: local, state, regional, national. Partners should include all levels of government, small to large businesses, academicians, scientists, chambers of commerce, nonprofits, trade associations, voluntary organizations, neighborhood groups, researchers, technical experts, professional societies, educators, and media, always leaving room for more at the partnership table. (See Figure 5.12.)

It is important to get broad representation, as participation and expertise are the unsung causes for Project Impact's success, with funding actually secondary. The reason for this is that if programs are only tied to funding, when the seed money is spent, the programs go away. In contrast, committed partners will work to institutionalize programs that become integrated into existing budgets or organizations, minimizing the need for large continuing outside funds. This is not to imply that money is anything less than necessary, but "in-kind" contributions will ultimately greatly outweigh direct funds in any sustainable effort.

"In-kind" funds are defined as any contribution where no direct funds are spent or the service is absorbed by the organization. Examples range from volunteer or staff time at an event to the use of in-house printers and other such resources that would otherwise require spending grant funds. "In-kind" contributions, together with some direct funds, provided Seattle Project Impact in just the first three years of its existence

FIGURE 5.12 A regular Seattle Project Impact Steering Committee meeting, but run at the 1998 Project Impact Summit in Washington, D.C., to demonstrate the collaborative partnership's inner workings and activities for the audience. (© FEMA)

$4.4 million in leveraged funds beyond the $1 million seed monies. In fact, this number is lower than actual as some partners did not want to track or publicize large in-kind contributions, considering them a "gift" to the effort. During that time, partners also volunteered more than 39,000 hours toward Seattle Project Impact, and the total of all partner contributions only grew exponentially in the six years following.

The next step undertaken was to create solutions for identified risks. In the earliest implementation stages, when beginning the tasks of identifying which programs would be focused on, an important lesson to all partners was to remember that a community's answers to mitigation requirements are best found at home. "Disasters are local,"[1] and therefore solutions to such issues must be locally found. Most communities both know their risks and have many initial ideas for solutions. However, if a community has no one with a particular skill-set in a needed area, such as a scientist or a researcher, bringing to bear outside expertise is critical to identifying or addressing hazards.

With Seattle OEM at the helm, partners provided leadership on the Seattle Project Impact Steering Committee, a task force with committees and subcommittees in charge of designing the detailed programs, including recruiting additional partners and subject-matter experts in order to enhance the development of overall community resilience. (See Sidebar 1.)

Sidebar I
Community Outreach

No matter how great a program is, if the public does not know anything about it, it serves no one. Among the Seattle Project Impact committees was one focused on outreach. We took advantage of every opportunity to keep mitigation in people's frame of reference.

We created the nationally acclaimed and comprehensive Seattle Project Impact Web site,[2] which provided great detail on each of the programs, resources, interactive games, photos of hazards, a kid's page, how-to information for mitigation and preparedness, many downloads, and a calendar that highlighted Home Retrofit classes, professional trainings, hazard workshops, clinics, and more.

- Another outreach tool was a public forum called "Disaster Saturday." This twice-a-year event was free to the public and allowed residents to get hands-on experience and have their questions answered by local experts. Workshops were held on a variety of mitigation and preparedness topics, displays and exhibits covered additional subjects, and there was always a giveaway to couple learning with fun. Home Retrofit was a big attraction, and visitors traveled from Canada and Oregon to attend.
- "Disaster Saturday" would generate thousands of attendees along with more to the smaller "Disaster Wednesday" evening alternative forums designed for those who may have had religious or other Saturday commitment. These and other community forums provided an opportunity to couple the education of home owners on their regional earthquake risk with specific information to help minimize its effects.
- Additional outreach projects included videos and televisions shows on specific programs, such as **Regional Home Retrofit.**
- Press releases and interviews with local, national, and international media.
- Multiple national awards of excellence and recognition.
- A multiyear Home Retrofit regional media campaign with several sponsors with both structural and nonstructural monthly in-store clinics preceded by radio spots and online information.
- Printed materials, such as lifeline hazard posters, brochures, fact sheets, the *Home Retrofit Series*, the *Nonstructural Protection Guide*, information packets, and hazard maps. (See Figure 5-14.)

FIGURE 5.13 Seattle Project Impact director Inés Pearce addressing the Disaster Mitigation Act of 2000's national "listening session" regarding the importance of public–private partnership in the Act. This came one month after Seattle experienced the magnitude 6.8 earthquake on February 28, 2001, and Seattle was asked to speak on behalf of all Project Impact communities. (© FEMA)

Committees met every two weeks, or more, and the Steering Committee met once a month, where overall goals and actions were created and additional insights or guidance was given to each program. The involvement of the partners in shaping all aspects of Seattle Project Impact created true collaboration, and therefore the program design occurred very rapidly, a great triumph. Due to partners' commitment, the vast majority of program design used no grant funds as time, ideas, and materials were contributed with the intent of long-term sustainability. Part of this came from individuals' excitement that they were involved in performing good for the community, and the vast majority had been never given such a groundbreaking opportunity to implement life-saving change. (See Figure 5.13.)

The Seattle Project Impact Steering Committee came up with a list of twenty possible ideas, but being a pilot community on a federal timeline for spending grant funds, we scaled the list down to three areas of focus that could be designed and implemented in two years while building in sustainability. Two would result in the reduction of the community's potential for disaster and one would result in improving the identification of natural hazards and assessment of vulnerability. The programs were

FIGURE 5.14 Attendees of a free "Disaster Saturday" event to learn more about Seattle Project Impact and neighborhood preparedness programs, attend workshops, visit booths, get hands-on instruction, and have questions answered. (© Seattle Project Impact)

Home Retrofit, School Retrofit, and better earthquake and landslide Hazard Mapping. Through these three programs we would be able to address the largest exposure to the city in its residences, protect children, and better understand our risk for future decision-making. (See Figure 5.15.)

The working relationships within the Steering Committee were excellent, and proved that the residential community, private sector, and public agencies can find common ground to work together proactively. FEMA continually held Seattle Project Impact in the national spotlight of how true partnerships work and can be successful in mitigating long-term risk reduction. (See Figure 5.16.)

Hazard Mapping

The geology of Seattle is directly responsible for many of the risks associated with natural disasters. The city has an extensive relief with large amounts of unconsolidated glacial soils. These geological features contribute landslides and intensified ground shaking from earthquakes, the city's two greatest hazards. The vision of the Hazard Mapping program

FIGURE 5.15 Seattle Project Impact Partner Award given to Phinney Neighborhood Association's Roger Faris, who taught thousands of home owners and contractors about Home Retrofit, and Ed Medeiros presented by previous winner Bank of America. Faris also received FEMA's "Outstanding Model Citizen Award" the previous year. (© Seattle Project Impact)

FIGURE 5.16 One of two national Outstanding Model Corporate Partner Awards in 1998 for Seattle Project Impact. FEMA Director James Lee Witt presented the award to Washington Mutual partner Doug Chandler.

FIGURE 5.17A Landslide on Seattle's Magnolia Hill in 1997, making houses vulnerable and damaging the bridge, which was closed for months, limiting access to that community. (© City of Seattle)

was to study geologic hazards and produce better landslide and earthquake hazard maps for public education and to use in developing sound land-use policy. This was a unique program as partners were most interested in being gathered under the Project Impact umbrella therefore able to work together cooperatively with other public and private partners, and the majority of this program did not use grant funds. Hazard Mapping became the foundation for all the other programs because if people do not understand their risks, they will see no need for action. Hazard Mapping fueled mitigation. A sign of success for increased hazard education was the increased public understanding that the Pacific Northwest is susceptible to three different types of earthquakes, which assisted residents in making better decisions about mitigation actions, such as retrofitting. (See Figure 5.17a and 5.17b.)

Seattle already had some hazard maps, but they needed to be updated. Under the Hazard Mapping umbrella, the U.S. Geological Survey (USGS) scientists, city Geographical Information Systems (GIS) staff, University of Washington (UW) academics and researchers, technical experts, and private firms worked together to share databases that improved risk information. Due to these partners' involvement, a landslide map was produced that integrated existing slide records beginning in the late-1800s with data about historical rainfall and the soil properties of Seattle's landslide-prone areas.

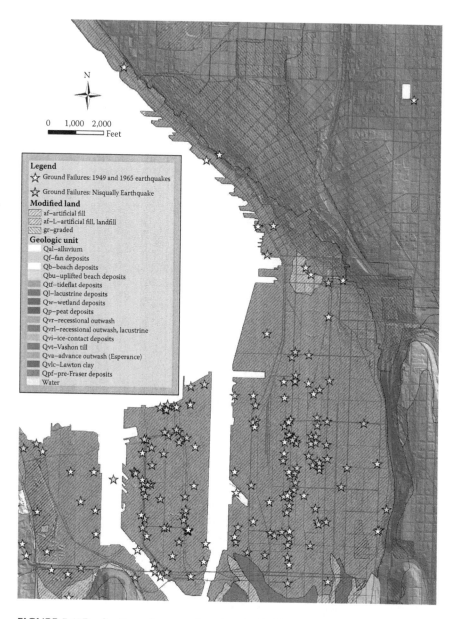

FIGURE 5.17B Section of a map showing drilled boring sites looking at below-surface soils, information that was incorporated into other maps to explain levels of ground shaking or likelihood of landslides. (© UW/USGS)

To address earthquakes, a three-dimensional geologic map of the Puget Sound Area was produced that incorporated complex relations beneath the surface with data of an earthquake's ground motion. While these two maps were expected for the program, ten other valuable projects spawned from all the collaboration and over ten years later, some are still continuing. The benefit of these expanded products is that difficult political choices are easier to sell to government officials when backed up by good science, which can be facilitated by new technology and information. (See Figure 5.18.)

One of the educational activities undertaken by partners was to provide "hazard tours" to elected officials, business leaders, city staff, and other decision makers with the purpose of showing them in-person where the hazards are, how to recognize them, and why they put people at risk. These tours changed the way leaders and residents saw their own city and region by making the hazard real and close enough to touch. The overall need for Hazard Mapping was continually demonstrated and committed to by partners, as there is always more to learn about the communities' risks. (See Figure 5.19.)

Home Retrofit

The vision of our second program, Home Retrofit, was to help home owners strengthen property susceptible to structural damage from earthquakes. In the city of Seattle, about 125,000 houses were built before the building codes changed in the late 1970s. When broadening the view to look at surrounding jurisdictions at risk in the region, that number jumps up to 250,000 potentially vulnerable homes. The issue with these wood-framed houses is due to three weaknesses:

1. They have no or insufficient bolts securing the frame to the concrete foundation and need to use square washers.
2. The cripple/sheer/pony walls are not reinforced laterally in order to withstand the side-to-side shaking.
3. The first floor is not connected to the foundation systems and eliminates the house responding as "one structure" from foundation to roof during shaking.

To address these vulnerabilities, Home Retrofit combined expertise with training, education with resources, and outreach with solution. It was a packaged program that focused on mitigation instead of waiting until after a disaster strikes to pick up the pieces. The impact of such a disaster

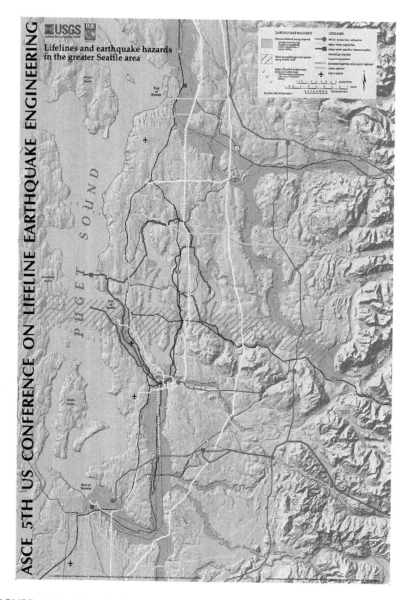

FIGURE 5.18 Map depicting critical regional lifelines, potentially liquefiable soils, the Seattle basin, and Seattle Fault; all when combined show expected areas for significant disruption. While many more detailed maps were generated, this was the first that galvanized community action (© USGS).

FIGURE 5.19 Seattle "Hazard Tour" given to the city building department, transportation, public works, and emergency staff led by University of Washington and USGS scientists. (© Seattle Project Impact)

can be minimized through retrofitting, be it structural or nonstructural, and Home Retrofit worked with all the contributors in the community: the homeowner, the contractor, the building officials, the public sector, neighborhoods, and the private sector. This provided local solutions to make a more disaster resilient region, one house at a time. (See Figure 5.20.)

"Regional Home Retrofit," or "Home Retrofit," is a comprehensive program with each element building on the next to ensure consistency of information, accuracy of retrofitting, and the protection of the home owner. This program began with planset drawings shared with us by San Leandro, California, for a program they had just started. Through the Home Retrofit Standards & Design subcommittee, these plansets were updated to depict the Pacific Northwest construction, which is similar to the California homes that sustained much damage in the 1989 Loma Prieta and 1994 Northridge Earthquakes. (See Figure 5.21.)

These pre-engineered Home Retrofit Plansets minimize the need for bringing in an engineer for a "standard" retrofit, therefore lessening the retrofit costs, and became the foundation for Home Retrofit. (See Sidebar 2.)

Although the design, implementation, and management of the Home Retrofit costs were covered by FEMA funds, the onus of actual costs for retrofitting was on the home owner. For those doing their own retrofit, using sweat equity, their cost would run $500 to $1,200, depending on the size of the house. For home owners hiring a contractor, the cost would

FIGURE 5.20   The Pacific Northwest has similar home construction to California, where this wood-frame house was damaged in the 1994 Northridge Earthquake. Square washers were added to retrofitting in response to the lessons learned in that earthquake.

vary from $3,000 to $7,000 for a standard retrofit. Since not everyone living in these earthquake vulnerable houses could afford the retrofit, we implemented a pilot grant program for low-to-moderate income homeowners.

At first the Home Retrofit Grant Program was run out of the Seattle Office of Housing, tied to their weatherization program and later it was run out of the Seattle Office of Human Services tied to their utility discount program. Through both of these efforts 187 homes were earthquake-strengthened using grants. But regardless of funding source, the fact is that this type of mitigation protects families, increases public safety, safeguards real estate investments, and fosters community recovery. This bore out after the 2001 Nisqually Earthquake when the majority of grant recipients shared their experience of and enthusiasm for a more disaster-resistant home with no damage, with only five reporting very minor damage. (See Figure 5.22.)

The participation of building departments was the key. For instance, the final inspection of any retrofit is important, regardless of whether a contractor or the home owner performed it, for the purpose of quality control. Home Retrofit prescriptive plans work with homeowners even if their home falls slightly outside of the standard plan. A building department can work with home owners to find reasonable, informed options to strengthen their house. With staff working to deliver solutions, the entire

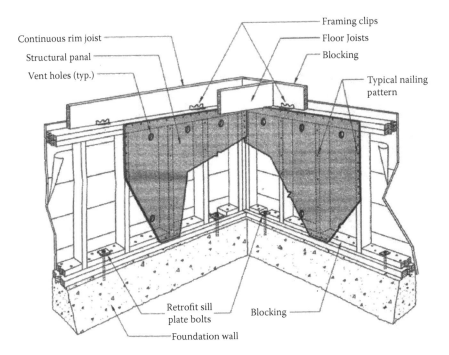

Continuous rim joist

Structural panal

Vent holes (typ.)

Framing clips

Floor Joists

Blocking

Typical nailing pattern

Retrofit sill plate bolts

Blocking

Foundation wall

Notes:
1. This sketch shows a sample wall section that has undergone a typical seismic strengthening retrofit.
2. This is a general sketch and is not intended to supersede requirements contained in the Standard Home Earthquake Retrofit Plan or in the specific installation details.

FIGURE 5.21 Regional Home Retrofit pre-engineered plans drawing showing the standard retrofit solution, used as the basis for home owner and professional training for quality control. (© Seattle Project Impact)

community benefits from each completed retrofit. Regional participation from building officials, engineers, plan examiners, architects, trainers, and contractors on the Standards & Design subcommittee ensured that once the program was up and running, it could be easily exported to other juris-dictions. Some jurisdictions also found grant funds for their low-income households. As twenty additional cities and counties launched Regional Home Retrofit, Seattle Project Impact continued to provide support to those building departments. The goal was to retrofit all vulnerable homes. By working together regionally on this effort we committed to protecting families through safer homes and safer communities. (See Figure 5.23.)

Sidebar 2
Home Retrofit Program Elements

1. Home Retrofit Plansets
2. Expedited, low-cost building-permit-driven process that established generic retrofit solutions for older, wood-framed homes and included two inspections
3. Two-hour home owner classes, a tool-lending library and technical assistance for home owners interested in doing their own retrofit, or consumer information for those home owners who would hire one of the trained contractors
4. Six-hour extensive training for contractors, building officials, architects, engineers, inspectors, and other building professionals through the University of Washington on these new retrofit standards after which contractors could choose to be included on a referral list for home owners
5. Special loan packages with partnering banks and a credit union

---

FIGURE 5.22 Yard sign in front of a Home Retrofit Grant recipient while the retrofit was in process. Many called after the 2001 earthquake to thank us for providing peace of mind to their family. (© Seattle Project Impact)

FIGURE 5.23   A full house in the always-popular Home Retrofit Class for home-owners taught by 2000 Outstanding Model Citizen Award–winner Roger Faris (right) and Kaveh Aminian (left) from the Seattle building department. (© Seattle Project Impact)

School Retrofit

The vision of our third program, School Retrofit, was to remove nonstructural and classroom hazards from schools to keep students, teachers, and staff safe. In 1949 and 1965 there were moderate earthquakes that caused significant structural damage to Seattle schools and also resulted in one student death. In the thirty years following the 1965 earthquake, voters passed $40 million in city referenda to upgrade Seattle schools. However, nonstructural retrofitting needs were not covered under these referenda. The Seattle Project Impact Steering Committee and Seattle Public Schools determined that a substantial earthquake risk remained from three categories of nonstructural elements, such as classroom hazards, nonflexible gas supply lines, and overhead hazards, such as large water tanks and radiators. (See Sidebar 3.)

This program focused on nonstructural classroom and building risks beginning with the update of the 1988 *Nonstructural Protection Guide,* which included a chapter on how to identify these risks and then how to mitigate by providing directions and images. This guide became the cornerstone for training to maintenance, custodial, and zone crews, who in turn would perform the more technical nonstructural retrofits.

Sidebar 3
Institutionalization

Mitigation cannot be successful if the programs are not designed for long-term solutions and made sustainable. Seattle Project Impact built institutionalization into its programs' design. For example, there are natural champions who already bear responsibility for certain aspects of what we began, such as School Retrofit. It would not be effective if Seattle Project Impact had been the lead on this effort. Instead, during the grant period we began to transition management to the Seattle Public Schools. Seattle Project Impact always provided committed support and assistance to keep the program moving as School Retrofit was the first of the programs to be institutionalized. An example of this was how when the school district performed light upgrades they also completed overhead retrofit work at the same time, which allowed for more cost effective approaches. Zone crews are now taking nonstructural mitigation measures as part of their every day activities. That is success — when mitigation is no longer thought of as an "extra" task on the to-do list. This was also the case with the Home Retrofit Training for Professionals, now managed by the University of Washington as a regular class in their extension program. (See Figure 5.24.)

Forty-six schools had overhead hazards removed, such as porcelain flush tanks and large water vats.

In other schools, teams secured valuable electronic equipment, including TVs, computers, and other items that could fall and pose a hazard to students by having shelving or equipment tied down, bolted, or reinforced. University of Washington students, partners, business representatives, volunteers, PTA members, and school and district staff members all participated in these half-day Saturday retrofits. School Retrofit also installed automatic gas shut-off valves to decrease the chance of earthquake-related natural gas leaks and performed an effectiveness study. (See Figure 5.25.)

As an example of the importance of School Retrofit efforts, during the February 28, 2001, Nisqually earthquake, one of the drained 300-gallon flush tanks had broken free of its restraints. If the tank had been full (weighing over 1 ton) and unretrofitted, the likelihood for extensive

FIGURE 5.24 School district staff securing library shelving in Asa Mercer Middle School. They now perform retrofit tasks as part of their daily duties. (© Seattle Project Impact)

injuries in the full classroom of children immediately below was very high. (See Figure 5.26.)

Seattle Project Impact took School Retrofit one step further by extending beyond the city limits to educate other districts that are susceptible to earthquakes. Other communities were impressed and lauded Seattle's presentations across the nation, with many more downloading

FIGURE 5.25   This University of Washington student participates on a School Retrofit team securing equipment in an elementary school computer lab. (© Seattle Project Impact)

the *Nonstructural Protection Guide* from the Seattle Project Impact Web site. Other communities took this lesson to heart and started similar programs using the model of Seattle Project Impact as demonstrated in Kenai, Alaska, and Walla Walla, Washington, which were two of the first to launch similar programs. The state of California is still using a version of the *Guide* today. (See Figure 5.27.)

In addition, Seattle Project Impact and Seattle Public Schools created the Student Education Campaign to Understand Risk from Earthquakes (SECURE) that involved students in creating options for making our community more earthquake-safe. In one middle school, students were selected winners of the Seattle Project Impact Poster Contest with their innovative ideas. At Nathan Hale High School, Seattle Project Impact, Seattle Public Schools, and FEMA sponsored the video *I Don't Fit Under My Desk: Advanced Earthquake Safety*, written by and for students. The high school students who created it won two regional accolades, including a Northwest Emmy Award. (See Figure 5.28.)

Business Disaster Mitigation

With the success of the three initial programs identified as most critical, a fourth area of concern became apparent. People in Seattle were becoming

FIGURE 5.26 The retrofitted, 300-gallon water vat in the attic of the Stevens at MacDonald Elementary School, which broke free of its restraints but had been drained, so no children were injured in the full classroom below the vat. (© Seattle Project Impact)

safer in their homes and their children safer in their schools, but what of the places that all working people spend at least a third of their work week throughout most of their adult lives — namely, work? Businesses are a vital part of the communities in which they operate.

In the event of even a small disaster, people who are safe in their homes and schools are still in need of food, prescription medication, gasoline, and other basic necessities. Recovery efforts can hardly begin in earnest if employees cannot return to work or if a business is so devastated that their operations were shutdown leaving their customers with no alternative. Imagine the recovery delays if an insurance company's claims adjusters were unable to perform their duties after a disaster. (See Figure 5.29.)

As the primary programs advanced, Seattle Project Impact added an unfunded program in 1999 to widen the mitigation scope. The Business

FIGURE 5.27 Seattle Public School's Theresa Salmon talks to elementary school children during an assembly about earthquakes and what to do when the earth shakes, a public education piece of the School Retrofit program. (© Seattle Project Impact)

FIGURE 5.28 A high school student showing a classmate how to Drop, Cover, and Hold on, in the Emmy-winning video *I Don't Fit under My Desk: Advanced Earthquake Safety* sold to generate funds for School Retrofit. (© Seattle Project Impact)

FIGURE 5.29 Damage to the Phoenix Underground in Seattle's Pioneer Square area that also impacted surrounding businesses as the road was closed for more than a year, limiting customer access.

Disaster Mitigation (BDM) program was developed to educate businesses about their disaster exposures and ways to minimize their economic impact in order to keep their doors open. Elements included a number of resources for all sizes of business: mentoring, short functional workshops, materials and Web links to recommended sites. While primarily targeting

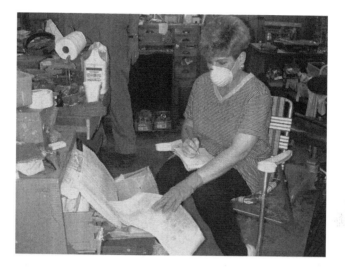

FIGURE 5.30 Example of an impacted business trying to recover paper files not backed up before a disaster, an issue shared with Seattle businesses that could cause major disruption and losses.

small- to medium-sized businesses, which are far and away the largest business segments in the economy, the information was provided to all businesses. (See Figure 5.30.)

From my years in the business continuity planning realm, I saw that there were issues not being addressed that could have serious implications on the community's ability to recover. Every community in the midst of a disaster, where the small businesses were hit hard, has experienced a delayed recovery time for the entire community. In some cases, those communities never fully recovered. As a community, we intersect our daily operations across sectors and industries. Impacts to one can affect the others. By working in advance to minimize businesses' exposure to loss of life and property, we would increase our community's resiliency. (See Figure 5-31.)

The BDM program began to compile various materials used for business planning and organize an effort to create different tools to assist businesses of all sizes. This effort sparked another one, wholly independent of but a full partner of Seattle Project Impact. Like BDM, it tried to create a guide referencing various source materials, but in the end found that it was not effective. The partners then took it upon themselves to write a tool from scratch that would truly meet businesses' needs. As in other areas where it is necessary to find natural integration of initiatives in

FIGURE 5.31 Damaged road closed having impacts on a business' employees, customers, suppliers, and vendors.

order to move it forward, this occurred with business mitigation. Seattle Project Impact was a major proponent of the effort brilliantly led by the Disaster Resistant Business (DRB) Toolkit Workgroup, a group consisting of mostly private business continuity experts and public planners.

The major product of this Workgroup's partnership was the development of the DRB Toolkit™. The Toolkit began shaping into a great mechanism that would help businesses and government by assisting businesses to take action, and could also aid Seattle Project Impact in not having to reinvent the wheel. As this tremendous effort was unfunded, the partners' participation to develop this first-ever tool was volunteer and therefore progressed steadily but more slowly than projects in the other programs. To further advance the availability of the DRB Toolkit, in 2006, the DRB Toolkit Workgroup became a nonprofit corporation, which will release the fruit of this collaborative as a tool that can assist Seattle Project Impact, all state and national partners who supported its development. (See Figure 5.32.)

CONCLUSION

To further support sustainability beginning with implementation, it was important to establish through the partnership and programs that the Project Impact concept must survive administrations. New partners,

FIGURE 5.32 Disaster Resistant Business (DRB) Toolkit Workgroup authors representing public and private organizations (left to right, John Ufford, Jim Hubly, Charles Davis Jr.) acknowledged for their multiyear voluntary effort to minimize business impacts.

emergency management staff and leadership, or elected officials should not shy away from Project Impact or like programs because it was not their idea. It makes business sense and is good government to focus on mitigation in this way. Even if this means that in order to safeguard the work, the new members of any management or political regime decide to change the name of the program in order to support it, new leadership can easily carry on the momentum; the benefits to them and the entire community are great.

A decision to end successful mitigation programs for political advantage or personal agenda is both reckless and unethical. One cannot imagine the logic employed by anyone who calls him- or herself a proponent of Emergency Management to justify walking away from one of the most successful efforts to truly make communities and the United States safer from disasters. When established programs only need continued support, any other action is unconscionable.

Because mitigation benefits are compelling in both human and economic terms, it was Seattle Project Impact's goal from the beginning that these efforts continue well into the future, and be shared with other communities facing the same risks. (See Sidebar 4.) (See Figure 5.33.)

Sidebar 4
Mentoring

It has always been a fundamental tenet of Seattle Project Impact's approach to community mitigation that we will be available to assist others to adapt successful Seattle programs to their respective community environments. Seattle Project Impact was recognized as a leader in promoting community mitigation throughout the country. Notable examples included working directly with the other 250 Project Impact communities by helping to develop the Disaster Resistant Communities Network (DRCN) with communities such as Oakland, California, and Tulsa, Oklahoma, then directly mentoring others. In sharing lessons learned, a valuable role of Project Impact was to help communities avoid reinventing the wheel. (See Figure 5.34.)

FIGURE 5.33 Seattle Project Impact's three-year anniversary to recognize partner contributions and program successes. This event had to be rescheduled due to the February 28 Nisqually earthquake, but was held three months later. (© Seattle Project Impact)

FIGURE 5.34 Seattle Project Impact shared lessons learned and programs one-on-one with communities or with many at community forums and seminars such as this one. (© Seattle Project Impact)

REFERENCES

1. Jim Mullen, Director, Washington State Emergency Management Division; Project Impact Summit, Washington D.C., December 1998.
2. Unfortunately, as of 2007, the comprehensive Seattle Project Impact Web site is no longer available, although a few pages were copied to the Seattle Emergency Management Web site.

6

Conclusions and Recommendations

George Haddow

George Haddow is a principal in the disaster management consulting firm of Bullock & Haddow LLC. He has worked on homeland security and emergency management projects with the Federal Emergency Management Agency (FEMA), the Corporation for National and Community Service, the Annie E. Casey Foundation, Save the Children, the Humane Society of the United States, the World Bank, and the Global Partnership for Preparedness. Mr. Haddow serves as an adjunct professor at the Institute for Crisis, Disaster and Risk Management at George Washington University in Washington, D.C. Mr. Haddow is the former deputy chief of staff to James Lee Witt during his tenure as Director of the U.S. Federal Emergency Management Agency (FEMA). At FEMA, Mr. Haddow was responsible for policy formulation in the areas of disaster response and recovery, public/private partnerships, public information, environmental protection, and disaster mitigation.

INTRODUCTION

The evidence is overwhelming that global warming will have a dramatic impact on future disasters. In coming years, the frequency and severity of disasters will continue to rise, and it is imperative that individuals and communities take action to reduce the impacts of these disasters.

Dramatically reducing the emissions that cause global warming is one form of hazard mitigation that must start immediately and continue unabated if we are to effectively eliminate global warming. However, it is also important that actions be taken to reduce the impact of future disasters caused or aggravated by global warming.

This book has attempted to address how we as a society can success-fully reduce the impacts of future disasters caused by global warming. We have presented a series of essays and case studies written by individu-als with real-world experience in dealing with the impacts of disasters. What we hoped to accomplish was to assure decision makers at all levels of society that reducing disaster impacts is not as daunting a task as it appears and that there are clear examples of how this can be successfully accomplished. We have provided information on the need for planners, environmentalists, and an entirely new set of professionals to join this effort, the various federal programs that can play a major role in hazard mitigation, and case studies from communities across the nation that have successfully designed and implemented hazard-mitigation programs that can serve as examples for the future.

The purpose of this chapter is to summarize the information pre-sented in the previous chapters and to highlight those factors that will be critical to dealing with climate change in the future. Also presented are a series of recommendations designed to guide the efforts of a full range of community stakeholders (i.e., local, state, and federal government, the private sector, nongovernmental organizations) in designing and imple-menting effective hazard mitigation programs to reduce the impacts of future disasters caused by global warming.

CONCLUSIONS: WHAT WORKS

The most important thing we have learned is that reducing disaster risks and losses is best done at the community level. Outside resources and technical assistance are critical, but effective and sustainable hazard mitigation programs and actions must be designed and implemented at the local level where disasters strike. Without the full support and participation of all stakeholders of a community in this effort it cannot be successful.

Based on the information presented in the essays and case studies in this book we conclude that the following factors are critical to building a successful community-based hazard-mitigation program:

- Involve all Community Stakeholders — Building a successful hazard-mitigation program cannot be done by a select few government officials. This work requires the involvement and efforts of all community stakeholders, including government officials (elected and appointed), emergency managers, first responders (fire, police, EMT), city and county managers, planners, community development officials, the local media, nonprofit groups, large and small business owners and associations, the Chamber of Commerce, environmental groups, developers, utilities, churches. Community organizations, voluntary and nongovernmental organizations, community-based organizations, community-base foundations and individual citizens.
- Local Champions — Leadership in this effort is all-important and a local champion can make the difference between success and failure. Such a champion can come from the public sector, the private sector, the nongovernmental sector, the community leadership, or everyday citizens. Carol Williams in Tulsa — an everyday citizen — was the first of many champions in that community's drive to reduce the impacts of flooding. The members of the Steering Committee for the International Flood Mitigation Initiative, which included two former governors, Canadian and U.S. government officials and leaders from the business and environmental communities were the initial champions for flood hazard mitigation in the Red River and successfully passed that mantle to the governors of North Dakota and Minnesota and the Premier of Manitoba. The University of California at Berkeley was a champion of earthquake hazard mitigation not only on its campus but in the community at large. The vineyard owners in Napa championed the community's 20-year flood-reduction plan and led the fight to secure a critical local funding source. Local champions come in all shapes and sizes and are all critical.
- Private Sector Involvement — The sustained health of any community is tied closely to the health of its economy and its private sector. In the aftermath of a disaster, the pace of recovery in the community can often be pegged to how quickly businesses and employers come back on line. Local business leaders and the local Chamber of Commerce must be involved in a community-based hazard mitigation effort. Their needs must be considered and the risks they face addressed. The private sector is an excellent source for local champions and one potential source for resources.

203

- Involving New Stakeholders — Hazard mitigation has long been the purview of emergency managers and operations. The time has come to expand the pool of government and nongovernment stakeholders beyond emergency management. Planners, community development specialists, and public administrators all need to be involved in hazard mitigation issues. Public interest groups, community-based organizations, and environmental groups all add to the expanse and effectiveness of efforts to reduce disaster risks and impacts. The new skills and relationships brought to the table by these new stakeholders will expand the reach and enhance the design and implementation of a community-based hazard mitigation program.
- Resources — As with any hazard mitigation program, resources will dictate much of the success of the program. Multiple sources of funding must be pursued including:
  - Local Funding Source — A local funding source can come in many forms, such as the city bond issues used in Berkeley, the ½-cent sale tax increase in Napa, and the storm-water drainage fee in Tulsa. A consistent and sustained local funding source provides reliable funding for a community-based effort and can be leveraged very effectively with funds from other government and non-government sources. Providing a local funding source sends a clear signal to the community-at-large and potential donors that a community is serious about hazard mitigation.
  - Federal and State Government Funds — The federal government is the most significant source for hazard mitigation funding. FEMA funds a variety of hazard mitigation programs, including its Hazard Mitigation Grant Program (HMGP) and the Disaster Hazard Mitigation Assistance (DMA) program. Other potential sources of Federal funding can be found in the various departments and agencies that have direct or indirect involvement in disasters, such as the Corps of Engineers, the Department of Transportation (DOT), Housing and Urban Development (HUD), Environmental Protection Agency (EPA), Health and Human Services (HHS), Department of Energy (DOE), and others. Nearly all government funding programs at the state and federal levels can be linked to hazard mitigation and with a little creativity accessed for use in a community-based hazard mitigation effort.

- Leverage Resources — The city of Seattle's Project Impact program was able to leverage its $1 million in seed money from FEMA with over $4 million in funding from other public and private sectors. Some of the resources were money and some in-kind donations but they all added up to Seattle being able expand its funding base fourfold. As noted earlier, a community can use a local funding source to leverage additional resources from other sources as well. Napa successfully used the funds from it ½-cent sales tax increase to match funds made available from state government and foundation programs. The key is to identify what sources of money and in-kind services are available among all potential sources including government, foundations, the private sector, and the nonprofit sector.
- Leadership from state and federal governments — Local champions are important but so are champions at the state and federal government levels. The support of President Clinton and his FEMA Director, James Lee Witt, served to validate local hazard mitigation efforts and to promote these programs with state and local decision makers, the public, and the media. FEMA's endorsement of hazard mitigation efforts in Tulsa and Berkeley through its Project Impact Community of the Year award allowed both communities to further expand and enhance their efforts. The involvement of the governors of South Dakota and Minnesota and the Premier of Manitoba have helped sustain the IFMI process and programs in the Red River basin. Leadership from state governors and the president play a significant role in the success of community-based hazard mitigation programs.
- Consensus Building — Coming to community consensus on risks and what can be done to mitigate those risks played a large role in the success of the programs presented in this book. Consensus building remains at the heart of the IFMI project in the Red River basin and was at the core of the 20-year flood-protection plan developed and implemented in Napa. These and other examples highlight how the consensus building process ensures that all stakeholders are involved and all views and ideas are heard. The resulting plans reflect the needs of all parties and include measures that can be supported and promoted by all involved.
- Environmental Protection and Enhancement — The critical role that a healthy and vibrant environment plays in reducing the

205

impacts of future disasters was clearly illustrated in the case studies involving flood issues. Restoring 900 acres of wetlands in Napa, creating additional open spaces and retention ponds in Tulsa, and creating the Greenway on the Red in the Red River basin are excellent examples of how the natural environment can be an effective ally in reducing flood impacts in a community.

RECOMMENDATIONS: HOW TO MAKE IT WORK IN YOUR COMMUNITY

Presented in the following sections are a series of recommendations on steps communities can take to design and implement a hazard mitigation program designed to reduce the impacts of future disasters. Recommendations on how the state and federal governments, the private sector, and the nonprofit sector can support communities in these efforts.

LOCAL GOVERNMENT

Hazard mitigation is best applied at the community level. If all politics are local, so, too, is hazard risk mitigation. In order to effectively reduce the impacts of future disasters caused by global warming, communities must make a commitment to understanding their risks and taking action. A community-based approach to hazard mitigation was successfully implemented in FEMA's Project Impact initiative in the late 1990s and has been recommended for communities hoping to adapt to the impacts of global warming by the Rockefeller Foundation and the ICLEI — Local Governments for Sustainability. (See Sidebar 1.)

The community-based approach is the spine of the hazard mitigation program designed to reduce the impacts of climate change that we propose to be supported by the actions of the state and federal governments, the business community, and the voluntary and nonprofit sectors.

This approach requires that communities take the following steps:

- Create a community partnership to lead the project that includes participation from all community members, including individual residents; government officials; the local Chamber of Commerce; large and small businesses; the local media, state, and local emergency management officials; police fire, and emergency medical

206

officials; community groups; churches; unions; nonprofits; environmental groups; social services; educational organizations, etc.

- Identify community and neighborhood hazard risks — What are the risks that your community faces (hurricanes, severe storms, tornadoes, flooding, drought, wildfire, etc.)? Which areas are vulnerable to these risks? Where are vulnerable and special needs populations located? Identify vulnerability of government facilities and the siting of emergency operations centers, etc. Identify the potential impacts on local businesses, schools, child care centers, homes, etc. Technical assistance should be provided by federal and state emergency-management officials.
- Identify and prioritize those actions, both structural and nonstructural, that can be taken by individuals, organizations, businesses, and the government in the community to lessen these risks and reduce the impacts of future disasters. (See Sidebar 2.)
- Communicate the plan to the community/neighborhood and generate the political, financial, and public support needed to implement the plan. Plan and conduct events and activities to promote and publicize the community's actions within the community and with national media, Congress, federal agencies, and other potential partners, etc. Organize support for sustaining community project efforts. Local media outlets can be very helpful in promoting the community partnership and the hazard mitigation efforts.

In support of these actions, undertaking the following activities is recommended:

- Establish Community Emergency Networks (CEN) designed to communicate hazard mitigation and preparedness messages to residents and to collect and provide information from residents to government, business, and nonprofit sector decision makers. Conduct a community-integrated demographic mapping project to identify hazard mitigation needs within the community, establish the CEN points of contact, and ensure participation among special needs groups, such as seniors, disabled, non-English-speaking, etc.
- Create a local funding source to provide the match for federal, state, and private funding for hazard-mitigation projects. Napa voters approved a ½-cent sales-tax increase to support their 20-year flood protection plan and for the past twenty years Tulsa has used a storm-water drainage fee to help fund critical flood-hazard mitigation

---

Sidebar 1
Examples of Community-Based Approaches

FEMA's Project Impact

"The goal of Project Impact is to bring communities together to take actions that prepare for — and protect themselves against — natural and manmade disasters in a collaborative effort. To accomplish this goal, we have organized pre-disaster activities into four phases:

1. Build a Community Partnership comprised of all community stakeholders.
2. Conduct a Hazard Identification and Hazard Vulnerability effort to examine the community's risks from natural and manmade hazards and to identify vulnerabilities to those risks.
3. Identify and Prioritize Risk Reduction Actions designed to mitigate identified risks and vulnerabilities and to reduce the impacts of future disasters.
4. Communicate your Plan to your Community in order to generate the public, political, and resource support needed to implement the Plan."

Source: Project Impact: Building a Disaster Resistant Community. Guidebook. Federal Emergency Management Agency, 1998.

ICLEI — Local Governments for Sustainability

"The purpose of *Preparing for Climate Change: A Guidebook for Local, Regional, and State Governments* is to help you as a decision-maker in a local, regional, or state government prepare for climate change by recommending a detailed, easy-to-understand process for climate change preparedness based on familiar resources and tools:

Scope the Climate Change Impacts to Your Major Sectors
Build and Maintain Support to Prepare for Climate Change
Build Your Climate Change Preparedness Team
Identify Your Planning Areas Relevant to Climate Change
Conduct a Climate Change Vulnerability Assessment
Conduct a Climate Change Risk Assessment
Set Preparedness Goals and Develop Your Preparedness Plan

Implement Your Preparedness Plan
Measure Your Progress and Update Your Plan"

Source: *Preparing for Climate Change: A Guidebook for Local, Regional, and State Governments.* Center for Science in the Earth System (The Climate Impacts Group), Joint Institute for the Study of the Atmosphere and Ocean University of Washington, King County, Washington, and ICLEI — Local Governments for Sustainability, 2007.

Judith Rodin: President of the Rockefeller Foundation

"The Rockefeller Foundation recently announced a major climate change initiative that concentrates on building resilience to a changing, challenging natural environment. As we see it, resilience incorporates five dimensions:

1. Information — effective adaptation will always be locally driven . . . communities need sophisticated measurement and assessment tools, integrated information about risks those tools reveal, and the best substantial approaches to minimize them.
2. Infrastructure — more than roads and sewers. It includes all the institutions and processes put in place to manage society.
3. Insurance — underserved populations need access to the social and economic security that comes from sharing risk. And the more people who share the risk, the lower the cost of coverage.
4. Institutional Capacity — Resilience requires that individuals and communities be empowered for and respond to crises from the ground-up. But there is also a critical role for governments and institutions to play in supporting resilience from the top down.
5. Integrated Systems — successful adaptation strategies . . . integrate urban planning, land-use regulation, water management, infrastructure investment, especially in energy and transportation, early-warning systems, and emergency and disaster preparedness, among many other elements."

Source: "Climate Change Adaptation: The Next Great Challenge for the Developing World." Remarks as delivered by Judith Rodin, president of the Rockefeller Foundation to the American Association for the Advancement of Science 2008 Annual Meeting. 2008.

Sidebar 2
Potential Hurricane and Flood Hazard Mitigation Actions

1. "Conduct audits of homes, child care centers, schools and neighborhood businesses to identify low cost actions that can be taken to reduce damage from future hurricanes and floods;
2. Restore and protect the natural environment to provide defense from storm surge and flooding including clearing streams of debris, restoring and protecting wetlands and creating open spaces in the community to soak up rain and flood waters;
3. Buyouts of properties in the floodplain;
4. Apply protective film to windows in schools, child care centers and low income and senior residences;
5. Evaluate all existing structural hazard mitigation entities such as levees, drainage and water diversion channels and flood protection gates."

Source: Project Impact: Building Disaster Resistant Communities Guidebook. FEMA, 1998.

Potential Drought Reduction Actions

1. "Connect regional water systems
2. Develop new groundwater sources
3. Implement new technologies such as reverse osmosis for desalinization
4. Provide financial incentives (e.g. tax breaks, rebates) for switching to more efficient manufacturing processes, irrigation practices and appliances
5. Renegotiate transboundary water agreements where applicable
6. Update drought management plans to recognize changing conditions
7. Increase authority to implement water restrictions and other emergency measures as needed
8. Expand use of climate information (e.g. seasonal forecasts) in water resources planning and management

9. Conduct additional research on how climate change may impact your community's water supply
10. Include information on climate change impacts to water supplies and how residents can reduce water use in utility inserts, newsletters, web sites, and local newspapers."

Source: Preparing for Climate Change: A Guidebook for Local, Regional, and State Governments. Center for Science in the Earth System (The Climate Impacts Group), Joint Institute for the Study of the Atmosphere and Ocean University of Washington, King County, Washington and ICLEI — Local Governments for Sustainability, 2007.

---

actions. In these and other communities across the country, locally generated funds have been used to match funding from federal and state governments, the business community, and the nonprofit and foundation communities to fund hazard mitigation actions.

- Establish an ongoing Monitoring and Evaluation Process that truly measures the benefits of the hazard mitigation actions to the community. Metrics should be established in each community to measure the reduction in disaster relief costs achieved by the hazard mitigation action, the economic benefits to the community of becoming more disaster resilient, and the multiple benefits realized from a healthy natural environment. This process will also evaluate the effectiveness of the community partnership and provide insights into how to improve all of its functions.

Creating community partnerships provides a direct benefit to the community and also provides collateral benefits to both federal and state governments, such as reduced disaster costs, stability of the tax base, continued economic development, and an overall increase in the health and safety of citizens, which leads to a more socially and economically healthy community.

FEDERAL SUPPORT FOR COMMUNITY-BASED HAZARD RISK MITIGATION

The federal government has a major role to play in promoting, designing, initiating, and funding programs and policies that will enhance community-based partnerships for hazard risk mitigation.

We are proposing that the federal government adopt a comprehensive hazard-mitigation strategy as one of its primary means for reducing the immediate impacts of global warming. We propose the following elements as essential to implementing a successful strategy:

- Establish an entity to serve as the federal focus for hazard-mitigation and long-term disaster recovery. Currently, limited federal hazard mitigation efforts are housed in the Hazard Mitigation Division within FEMA within DHS. In light of the responsibilities of DHS, it is understandable that the issues of hazard mitigation are not high on the department's agenda. FEMA is only focused on responding to the next disaster. In the aftermath of Katrina, a long-term recovery czar was named, but, as we have seen with the unacceptable level of progress in New Orleans, this position has little clout at present. In support of the federal government's role in advancing hazard mitigation as a primary step in addressing global warming, we need to establish an organization at the federal level that has advancing that goal as its primary mission. The entity must provide visible leadership, be flexible, and not create a new bureaucracy, be adaptable to changing technologies, and be transparent and accountable. One of the models to be considered would be an organization similar to the Appalachian Regional Commission, an organization with a limited mission and designed to address a specific need. To support the mission of the organization, the entity would exercise the following functions:
  - Administer a national fund to promote and financially support hazard mitigation activities to reduce the impacts of global warming and post-disaster recovery actions to ensure hazard mitigation is included.
  - Act as a clearinghouse and dissemination point for information on innovative strategies/best practices in hazard mitigation and reducing the impacts of global warming.
  - Develop partnerships with universities and the private sector to support problems-focused research to identify new strategies and technologies, especially for emerging hazards such as drought and urban wildfires.
  - Provide incentives to the private sector to incorporate hazard mitigation into economic development projects and infrastructure development.

- Work with Congress to revise existing federal disaster legislation and policies to incorporate disaster-resistant construction in public buildings, facilities, and infrastructure. Current interpretation of disaster policy is that the federal government supports rebuilding facilities to their condition pre-disaster. This means that federal dollars could end up rebuilding a hospital or a school in a coastal area impacted by a hurricane year after year because the local government alone cannot afford the cost of making the building disaster resilient. This is shortsighted on the part of the federal government for numerous reasons including potential additional costs to the National economy for delayed recovery within that community or region. It is in the best financial interests of the federal government to support "building back better."
- Review and revise capital and infrastructure funding programs of the federal government to incorporate and fund hazard mitigation in the design and construction of projects. An effort during the late 1980s resulted in changes to the Intermodal Surface Transportation Act that required all federally funded projects to assess the earthquake threat and incorporate earthquake resistant construction. This could be the model to look at other hazards, particularly the wind hazard from hurricanes (and tornadoes and blast). Certain wind criteria is incorporated into building codes, but many federal facilities and projects are not required to adhere to state or local building code provisions. The federal government should be the leader in disaster-resistant construction and not exercise the ability to be an allowable exception to the codes.
- Revise federal mapping efforts to incorporate indications of already evidenced impacts of global warming and opportunity areas for hazard mitigation. For example, one of the largest federal mapping efforts is the National Flood Insurance Program (NFIP) community maps. These maps, when revised, try to incorporate the latest development trends, but they are not juxtaposed against potential hurricane inundation or potential wind-speed impact areas, which would help in the development of revised building codes and standards and in evacuation planning. Other federal mapping programs, such as the U.S. Geological Survey, should be examined to see if there are other applications that would be beneficial, such as soils stability and liquefaction for disaster-resistant construction.

- Reinvigorate the climate threat program initiated by the Environmental Protection Agency (EPA). This program was initiated by
  the EPA in the 1990s to assist regions and states to better understand the climatologically changes and threats that they are facing.
  The program has languished under the Bush Administration. As
  noted earlier, climate change is contributing to all weather-related
  natural disasters, as clearly evidenced not only by the severity and
  frequency of hurricanes, tornadoes, and winter storms, but also in
  the dramatic increase in wildfires throughout the United States
  and, indeed, the whole planet.
- Explore the feasibility of a federally backed catastrophic all-hazards
  insurance program that requires hazard mitigation as a quid pro
  quo for the federal backing. The 2004 Florida hurricanes and the
  2005 Katrina and Rita hurricanes raised issues as to the liquidity
  of the private insurance markets. It also raised serious questions
  as to the fairness or equality of the coverage. For example, some
  insurers covered wind-driven rain from the storms and other
  did not. This also became an issue for coverage under Federal
  NFIP policies. Citizens in non-hurricane-prone areas felt they
  were unfairly paying for the conscious decisions people made
  to live in a dangerous area. This same argument has been heard
  about subsidizing people who live in earthquake-prone areas.
  An independent review of the need for and feasibility of a catastrophic all-hazards program would be beneficial to better understand the dynamics currently in the marketplace. Determining
  the value of such a program in advancing the goals of hazard
  mitigation would be an essential component of any study and
  recommendations.

What we are proposing is a very aggressive strategy that will achieve
significant long-term reductions in the impacts of future disasters in the
United States. It is a major step toward dealing with the problems presented
by climatological changes brought on by global warming. We believe it is
essential for the federal government to provide the leadership to implement this strategy. The question then remains how the federal government
will pay for this aggressive strategy. Several options are worth exploring:

Option 1. Balance pre-disaster hazard mitigation with post-disaster
costs. Determine how much we are currently spending on post-
disaster costs; perhaps look at the average of the past several

years (approx. $2.5 billion/year) and ask for a one-time supplemental, funded the same way disasters are funded. This would be off-budget, as is disaster funding, with the argument being that it will reduce future losses, which could easily be documented. A portion of this funding would be made available for reinvestment to provide an ongoing capitalization of the elements of the strategy.

Option 2. Create a Hazard Mitigation Trust Fund. A fee is added to every contract for building or upgrading any facility or project supported by federal funding (the amount could be as low as $10 dollars or as high as $1,000 — a sliding scale could be created based on the cost of the project or the use of the facility — schools could be less). This may be reasonable but the costs would end up coming back to the government unless it was written into the contract as a contractor contribution. Another option under this category would be to add $1 to $1000 to every building permit for development, the size of the addition based on the size of the development, but since permitting is a local issue there may be some resistance from local governments.

Option 3. Create a tax check-off for hazard-mitigation efforts, similar to the tax check-offs for energy.

Option 4. In looking at the all hazards insurance program, design the program and rates to allow for collection of funds to support hazard mitigation.

Option 5. Create a Hazard Mitigation Investment Bank. This would be an entity that would be supported by those private sector industries that benefit most from the investment in reducing disaster impacts, i.e., mortgage bankers, building and construction industries, architects, utility companies, etc. In exchange for investment, clients would get some level of tax incentive or other incentive. It could be designed so they could get some level of return on their investment.

Option 6. Include a risk cost in the loan percentages of any federal construction loan, or any federal backed or purchased mortgage. Properties covered under the financial instruments would be eligible to apply for hazard-mitigation grants from the fund.

These options are not independent of each other, and it is likely that implementation of some combination of the options would be necessary and attractive as the strategy moves forward.

STATE GOVERNMENT SUPPORT FOR
COMMUNITY-BASED HAZARD RISK MITIGATION

Global warming has been acknowledged and recognized as having a significant impact on the economies and resiliency of state operations. Because of the absence of federal leadership, we have seen states and governors take very proactive positions and provide leadership to address the impacts of global warming. Actions they have advanced have addressed the larger issues of global warming. Our proposals to address the immediate impacts of the increased severity and frequency of disasters as a result of global warming seek to address a more immediate problem faced by those states across the nation.

- Incorporate hazard mitigation elements in all state-supported building and infrastructure construction and require a percentage (somewhere between 10 to 20 percent) of this construction to be "green" construction. This action can result in safer buildings and reduced state disaster costs in the aftermath of a hurricane. This is an important consideration, since many states are self-insured and a large disaster can represent significant budget problems for a state even if eligible for federal assistance. Simple measures, such as incorporating protective film on windows in schools, courthouses, administrative buildings, etc., dramatically reduce damages and have the collateral benefit of improving energy efficiency. Other ideas include consideration of risk in siting of facilities and design of facilities to deal with hurricane wind levels. The costs of incorporating hazard reduction into new construction are minimal and are offset by potential losses.
- Develop and support state university teams to provide technical assistance to communities in reducing the impact of global warming through hazard mitigation. Public/community service has become a very positive requirement in many university curricula. Within schools of engineering, architecture, planning, public administration, and other related disciplines, part of the public and community service opportunity could be in assisting communities to design and implement hazard mitigation strategies for the community, for particular elements of the community, or even for individual buildings and structures. Universities could support problem-focused research to identify new and improved hazard-mitigation techniques to reduce disaster risks and impacts.

216

- Provide state tax incentives to private sector to incorporate hazard mitigation in their facilities and capital improvement programs. Revise economic incentives for business development to provide a premium for incorporation of hazard mitigation in siting and construction of new facilities or retrofit of existing facilities. Two of the most effective tools that states have at their disposal are their ability to create tax or economic incentives for supporting or promoting improved construction and development practices. Introducing the concepts of hazard mitigation as cost-effective and energy and environmentally friendly at the state level, makes sense and complements local processes where the incentive process is widely accepted and understood. Green spaces and public areas have long been part of the incentive package and hazard mitigation alternatives that complement these approaches are just logical. States can easily support the need for these incentives with public safety and cost-savings arguments as there is data to support offering these types of incentives. Statistics indicate between 40 to 60 percent of small businesses never recover from a disaster. States can support business adoption of hazard-mitigation incentives by providing technical assistance and education to businesses on the post-disaster economic benefits of reducing risks to cut losses.
- Revise state-wide building codes to require cost-effective efforts to reduce disaster impacts in all new construction and in any level of remodeling or reconstruction that impacts 25 percent or more of the building. In the United States, most building codes are established at the state level with many states allowing for adoption of local codes. Injecting disaster-impact reduction into new construction is the most cost-effective way to implement disaster-resilience actions and, in many cases, can be used as an effective marketing tool. The states with support from the federal government should identify a menu of the most effective and cost-effective hurricane hazard-mitigation measures for a variety of construction types. These need to be incorporated into state building codes. Second, most states require upgrades to current code based on a 50 percent or greater impact on the building. Consequently, many buildings come in assessed as only 49 percent impacted. If this threshold was lowered to 25 percent, almost all major renovations would be impacted and a much higher level of disaster resilience would be achieved.

217

- Review jobs programs administered at the state level with support from the Department of Commerce (DOC) and the Economic Development Administration (EDA) that focus on ensuring disaster-resistant jobs so people don't lose their jobs in the event of disaster.

BUSINESS COMMUNITY SUPPORT FOR COMMUNITY-BASED HAZARD RISK MITIGATION

Actions that the business community could conduct in support of community hazard risk mitigation efforts include:

- Provide leadership at the national, state, and local levels for hazard mitigation efforts. Community champions are a critical element in the success of any community-based effort, and often the most effective champions come from the business community. National organizations such as the U.S. Chamber of Commerce, the Business Roundtable, the National Association of Home Builders, the Realtors, and others should work to promote these community hazard-mitigation efforts and to encourage their members to take a leadership role at the community level.
- Become full members of the community partnership established in each community. Hundreds of corporations became Project Impact national partners. In Tulsa alone 345 local businesses joined Tulsa's Project Impact program.
- Provide financial, material (i.e., products, services, etc.), and technical support to community efforts. Across the country in the late 1990s, major corporations and local businesses provided funding and materials to support community hazard mitigation efforts.
- Allow skilled employees to take paid leave to assist community partnerships and to help implement hazard-mitigation actions. Major employers in a community could allow their risk managers and business continuity planners to help small businesses identify low-cost hazard-mitigation actions they can take to protect their business, local businesses could provide computer specialists to help with community mapping projects, and local businesses could contribute the services of their construction supervisors and employees to retrofit homes, child care centers, and small businesses, etc.

- Take steps to reduce the impacts of global warming on their operations, facilities, and employees. Complete business impact analysis for all facilities and operations; assist vendors and suppliers in doing the same. Provide grants and low-interest loans to employees to finance low-cost hazard mitigation actions to protect their homes.

- Include hazard mitigation, energy conservation, and environmentally friendly techniques in siting, designing, and constructing future facilities and retrofitting existing facilities — make business facilities and operations the model for all community members to follow in ensuring that the community's economy is well protected from future hurricanes.

NONPROFIT SECTOR SUPPORT FOR COMMUNITY-BASED HAZARD RISK MITIGATION

The voluntary agencies active in disasters (VOADs) have long played a major role in responding to disasters. These groups provide immediate food, shelter, and clothing to individuals and families impacted by hurricanes. They also help communities to rebuild after these events.

In recent years, the VOADs have been joined by an increasing number of nongovernmental organizations (NGOs) that provide funding, staff support, and partnerships to help individuals and communities to recover. Corporate, family, and community foundations have also become more involved in response and recovery efforts to major disasters.

It is time that these entities support hazard mitigation efforts in communities across the country. Actions they might take include:

- Provide trained and experienced staff to help organize community partnerships, to design community hazard mitigation action plans, and to implement these plans. The non-profit sector has volunteers and paid staff located in communities across the country who could become involved in community hazard mitigation efforts by adding this function to existing community activities and/or creating new programming to support hazard mitigation efforts. The relationships already established by these groups in communities could be leveraged very effectively to support a new community disaster-resistant initiative.

- Provide financial and technical support for the development of training and mentoring materials to provide guidance to new communities. Foundations could provide the funding, and community-based organizations and programs could provide the expertise needed to develop training and mentoring programs and to make this these programs available to community leaders around the country.
- Provide financial support for staffing for the community partnerships. The Council on Foundations has published a report entitled "Reducing the Impacts of Disasters on Children: Opportunities for Foundations," which provides a list of 35 activities that foundations could invest in that would help reduce the impacts of future disasters on children. The council should produce a similar report for community-based hazard mitigation efforts and distribute and promote it to its members. Such a report could serve as a guide for how individual foundations could craft new funding programs to support community hazard-mitigation efforts.
- Provide financial and technical support for the establishment of a Monitoring and Evaluation (M&E) Program — Non-profit groups have extensive experience in designing and implementing M&E Programs. This expertise should be used to design and implement the types of M&E programs needed to measure progress in community hazard mitigation efforts, measuring economic benefits realized through risk and impact reduction, and to determine the savings realized by reduced losses from future disasters.
- Take steps to reduce the impacts of global warming on their programs, operations, and facilities. Just as the business community should take steps to reduce future impacts, so should the non-profit sector. As this sector becomes more and more involved in disaster response and becomes more critical to the successful recovery of communities from disasters, it is important that their operations, services, and facilities remain functional after a disaster, and conducting an audit of these operations and facilities and taking action to reduce risk will ensure that they are better able to serve when needed.
- Incorporate hazard mitigation planning and actions into their existing community development programming — The non-profit sector supports a myriad of development programs in low-income and disadvantaged neighborhoods and communities. These are the areas and populations that are often the hardest hit by

disasters. It is critical that the non-profit sector understand the risks from disasters that their development programs face and take action to reduce these risks so that damage to these programs is limited when the next hurricane strikes.

- Ensure that all special needs populations are represented in the community partnership and the needs of these populations considered in all planning and design functions. The non-profit sector has a long history of working with special needs populations and should take an active role in ensuring that their needs are recognized and considered in all hazard mitigation efforts. The non-profit sector must leverage its ongoing work with special needs populations to ensure their involvement in the community hazard-mitigation efforts.

CONCLUSION

There are solutions to reducing the impact of the changing climate. Many of them are based on proven, effective, and cost-efficient activities that communities across the country have already taken to reduce the impact of future disasters. These lessons and these processes must be applied now and urgently, given what we know.

APPENDIX:
COMPILATION OF REPORTS, WEB SITES, AND OTHER MATERIALS RELATED TO CLIMATE CHANGE

Damon P. Coppola

Damon P. Coppola is author of several emergency management academic and professional texts, including *Introduction to International Disaster Management*, *Introduction to Homeland Security*, and *Introduction to Emergency Management*. He is also co-author of two FEMA Emergency Management Institute publications, *Hazards Risk Management* and *Emergency Management Case Studies*. As senior associate with the Washington, D.C.–based emergency management consulting firm Bullock & Haddow, LLC, Mr. Coppola has provided planning and technical assistance to emergency-management organizations at the local, state, national, and international levels, and in both the nonprofit and private sectors. Mr. Coppola received his master's in Engineering Management (MEM) degree in crisis, disaster, and risk management from the George Washington University.

---

* Disclaimer: The following resources are provided for informational purposes only. They do not necessarily represent the views of the authors or the publisher, and they do not represent any form of endorsement of their content by either the authors or the publisher.

REPORTS

- *Climate Change 2007 — The Physical Science Basis*
  Contribution of Working Group I to the Fourth Assessment
  Report of the IPCC
  Intergovernmental Panel on Climate Change (IPCC), 2007.
  http://www.ipcc.ch/ipccreports/ar4-wg1.htm
  Excerpt: "Representing the first major global assessment of climate
  change science in six years, 'Climate Change 2007 — The Physical
  Science Basis' has quickly captured the attention of both policymakers and the general public. The report confirms that our
  scientific understanding of the climate system and its sensitivity
  to greenhouse gas emissions is now richer and deeper than ever
  before. It also portrays a dynamic research sector that will provide
  ever greater insights into climate change over the coming years.
  The chapters forming the bulk of this report describe scientists'
  assessment of the state-of-knowledge in their respective fields. They
  were written by 152 coordinating lead authors and lead authors
  from over 30 countries and reviewed by over 600 experts."
- *Climate Change 2007 — Impacts, Adaptation and Vulnerability*
  Contribution of Working Group II to the Fourth Assessment
  Report of the IPCC
  Intergovernmental Panel on Climate Change (IPCC), 2007
  http://www.ipcc.ch/pdf/assessment-report/ar4/wg2/ar4-wg2-
  intro.pdf
  Summary: *Climate Change 2007: Impacts, Adaptation and Vulnerability*
  is the second volume of the IPCC Fourth Assessment Report. After
  confirming in the first volume on 'The Physical Science Basis' that
  climate change is occurring now, mostly as a result of human activities, this volume illustrates the impacts of global warming already
  under way and the potential for adaptation to reduce the vulnerability to, and risks of climate change. Drawing on over 29,000 data
  series, the current report provides a much broader set of evidence
  of observed impacts coming from the large number of field studies
  developed over recent years. The analysis of current and projected
  impacts is then carried out sector by sector in dedicated chapters.
  The report pays great attention to regional impacts and adaptation
  strategies, identifying the most vulnerable areas. A final section
  provides an overview of the inter-relationship between adaptation
  and mitigation in the context of sustainable development."

- *Climate Change 2007: Mitigation of Climate Change
  Contribution of Working Group III to the Fourth Assessment
  Report of the IPCC*
  Intergovernmental Panel on Climate Change (IPCC), 2007
  http://www.ipcc.ch/pdf/assessment-report/ar4/wg3/ar4-wg3-
  frontmatter.pdf
  Summary: "In the first two volumes of the 'Climate Change 2007'
  Assessment Report, the IPCC analyses the physical science basis
  of climate change and the expected consequences for natural
  and human systems. The third volume of the report presents an
  analysis of costs, policies and technologies that could be used to
  limit and/or prevent emissions of greenhouse gases, along with a
  range of activities to remove these gases from the atmosphere. It
  recognizes that a portfolio of adaptation and mitigation actions is
  required to reduce the risks of climate change. It also has broad-
  ened the assessment to include the relationship between sustainable development and climate change mitigation."
- *Climate Change 2007: Synthesis Report, Summary for Policymakers*
  **Intergovernmental Panel on Climate Change (IPCC), November
  2007**
  http://www.ipcc.ch/pdf/assessment-report/ar4/syr/ar4_syr_
  spm.pdf
  Summary: "This Synthesis Report is based on the assessment
  carried out by the three Working Groups of the Intergovernmental
  Panel on Climate Change (IPCC). It provides an integrated view of
  climate change as the final part of the IPCC's Fourth Assessment
  Report (AR4). A complete elaboration of the Topics covered in this
  summary can be found in this Synthesis Report and in the underlying reports of the three Working Groups."
- *Preparing for Climate Change: A Guidebook for Local, Regional
  and State Governments*
  **ICLEI: Local Governments for Sustainability, Center for Science
  in the Earth Systems (The Climate Impacts Group), Joint
  Institute for the Study of the Atmosphere and Ocean, University
  of Washington and Kong County, WA, September 2007**
  http://www.iclei.org/index.php?id=7066
  Excerpt: "Public decision-makers have a critical opportunity —
  and a need — to start preparing today for the impacts of climate
  change, even as we collectively continue the important work of
  reducing current and future greenhouse gas emissions. If we wait

225

until climate change impacts are clear to develop preparedness plans, we risk being poorly equipped to manage the economic and ecological consequences, and to take advantage of any potential benefits. Preparing for climate change is not a 'one size fits all' process. Just as the impacts of climate change will vary from place to place, the combination of institutions and legal and political tools available to public decision-makers are unique from region to region. Preparedness actions will need to be tailored to the circumstances of different communities. It is therefore necessary that local, regional, and state government decision-makers take an active role in preparing for climate change, because it is in their jurisdictions that climate change impacts are felt and understood most clearly."

- *A Survey of Climate Adaptation Planning*
  The H. John Heinz III Center for Science, Economics and the Environment, October 2007
  http://www.usecosystems.org/NEW_WEB/PDF/Adaptation_Report_October_10_2007.pdf
  Excerpt: "As evidence accumulates that a warming planet will cause widespread and mostly harmful effects, scientists and policy makers have proposed various mitigation strategies that might reduce the rate of climate change. For those officials in government who must plan now for an uncertain future, however, strategies for adapting to climate change are equally important. The options available to planning officials have become better defined over time as they have been studied — and in some cases, implemented — but adaptation planning continues to involve many uncertainties. These arise from the fact that every community is unique in its setting and people, and therefore faces environmental and social vulnerabilities that will differ from those of neighboring communities. Understanding the nature of these vulnerabilities is part of the challenge of creating an adaptation strategy."
- *Adaptation and Vulnerability to Climate Change: The Role of the Finance Sector*
  CEO Briefing: UNEP FI Climate Change Working Group (CCWG), November 2006
  http://www.unepfi.org/fileadmin/documents/CEO_briefing_adaptation_vulnerability_2006.pdf
  Excerpt: "Climate change is now certain, so we must plan for the reality that dangerous changes in weather patterns will disrupt

economic activity. On one scenario, disaster losses could reach over 1 trillion USD in a single year by 2040. The impacts will be worse in developing countries, where capacity to manage disasters is lower, and could impede progress towards achieving the Millennium Development Goals. Adaptation — adjusting to the expected effects of climate change — is therefore a clear imperative and a vital complement to mitigation. At the same time, a new integrated approach is called for to optimize the response of key actors in business, government and civil society. Such an approach should coordinate adaptation, disaster management, and sustainable economic development more systematically. Already the financial sector is incurring additional costs from adverse climatic conditions, and has developed and refined important techniques to cope with these burdens. The sector is restricted, however, by commercial considerations from applying these measures more widely. A gathering weight of opinion suggests that a combined public-private approach to adaptation could yield worthwhile results. Inevitably, returns would be small to begin with, but could grow rapidly as best practice spreads."

- *Weathering the Storms: Options for Framing Adaptation and Development*
  World Resources Institute, 2007
  http://www.wri.org/publication/weathering-the-storm#
  Synopsis: Clarifies the relationship between adaptation and development by analyzing 135 projects, policies, and other initiatives from the developing world that have been labeled by implementers or researchers as "adaptation to climate change."
- *King County 2007 Climate Plan*
  King County Government, WA, February 2007
  http://www.metrokc.gov/exec/news/2007/pdf/climateplan.pdf
  Excerpt: "(The plan) provides an overview of how King County seeks to reduce greenhouse gas emissions and works to anticipate and adapt to projected climate change impacts, based on best available science; it sets a process in motion to embed climate change mitigation and adaptation as critical factors in the cost-benefit evaluations of all decisions made by King County; it is a companion plan to the 2007 King County Energy Plan, a document detailing internal policies, programs and investments in climate-friendly, renewable energy that are critical to reducing operational greenhouse gas emissions and reducing dependence

on foreign fossil fuels; and it builds on over 15 years of efforts across King County departments to stop the causes of climate change and to prepare for regional climate change impacts. King County has taken significant steps in the past to address climate change. Nevertheless, this is the first document that brings all of King County's actions related to climate change together in one single plan."

- *A Climate Risk Management Approach to Disaster Reduction and Adaptation to Climate Change*
  United Nations Development Programme, 2002
  http://mona.uwi.edu/cardin/virtual_library/docs/1140/1140.pdf
  Summary: "UNDP report on new concept of 'Integrated Climate Risk Management' — integrating disaster reduction with adaptation to climate change and sustainable development imperatives."

- *Adapting to Climate Change: What's Needed in Poor Countries and Who Should Pay*
  Oxfam International, 2007
  http://www.oxfam.org.uk/resources/policy/climate_change/downloads/bp104_adapting_to_climate_change.pdf
  Summary: "Climate change is forcing vulnerable communities in poor countries to adapt to unprecedented climate stress. Rich countries, primarily responsible for creating the problem, must *stop harming*, by fast cutting their greenhouse-gas emissions, and *start helping*, by providing finance for adaptation. In developing countries Oxfam estimates that adaptation will cost at least $50bn each year, and far more if global emissions are not cut rapidly. Urgent work is necessary to gain a more accurate picture of the costs to the poor. According to Oxfam's new Adaptation Financing Index, the USA, European Union, Japan, Canada, and Australia should contribute over 95 per cent of the finance needed. This finance must not be counted towards meeting the UN-agreed target of 0.7 per cent for aid. Rich countries are planning multi-billion dollar adaptation measures at home, but to date they have delivered just $48m to international funds for least-developed country adaptation, and have counted it as aid: an unacceptable inequity in global responses to climate change."

- *An Abrupt Climate Change Scenario and Its Implications for United States National Security*
  Peter Schwartz and Doug Randall, 2003
  http://www.grist.org/pdf/AbruptClimateChange2003.pdf

Excerpt: "There is substantial evidence to indicate that significant global warming will occur during the 21st century. Because changes have been gradual so far, and are projected to be similarly gradual in the future, the effects of global warming have the potential to be manageable for most nations. Recent research, however, suggests that there is a possibility that this gradual global warming could lead to a relatively abrupt slowing of the ocean's thermohaline conveyor, which could lead to harsher winter weather conditions, sharply reduced soil moisture, and more intense winds in certain regions that currently provide a significant fraction of the world's food production. With inadequate preparation, the result could be a significant drop in the human carrying capacity of the Earth's environment. The research suggests that once temperature rises above some threshold, adverse weather conditions could develop relatively abruptly, with persistent changes in the atmospheric circulation causing drops in some regions of 5–10 degrees Fahrenheit in a single decade. Paleoclimatic evidence suggests that altered climatic patterns could last for as much as a century, as they did when the ocean conveyor collapsed 8,200 years ago, or, at the extreme, could last as long as 1,000 years as they did during the Younger Dryas, which began about 12,700 years ago."

- *Beyond the Ivory Tower: The Scientific Consensus on Climate Change*
  Naomi Oreskes, in *Science*, vol. 306, no. 5702 (2004), 1686
  http://www.sciencemag.org/cgi/content/full/306/5702/1686
  Excerpt: "Policy-makers and the media, particularly in the United States, frequently assert that climate science is highly uncertain. Some have used this as an argument against adopting strong measures to reduce greenhouse gas emissions. For example, while discussing a major U.S. Environmental Protection Agency report on the risks of climate change, then-EPA administrator Christine Whitman argued, 'As [the report] went through review, there was less consensus on the science and conclusions on climate change.' Some corporations whose revenues might be adversely affected by controls on carbon dioxide emissions have also alleged major uncertainties in the science. Such statements suggest that there might be substantive disagreement in the scientific community about the reality of anthropogenic climate change. This is not the case."

- *Catalyzing Commitment on Climate Change: A Paper on the International Climate Change Taskforce*
  Retallack, Simon, and Tony Grayling; Institute for Public Policy Research, 2005
  http://ippr.nvisage.uk.com/ecomm/files/Catalysing%20commitment.pdf
  Excerpt: "The negotiation of the UN Framework Convention on Climate Change and the Kyoto Protocol constitutes a major political achievement. Without Kyoto, there would be further delays in reducing emissions that could result in irreversible damage to the climate system. However, greenhouse gas emissions continue to rise. The key weakness of the international regime lies in its inability to gain traction: governments have so far failed to ensure that climate objectives are integrated in key policy areas, such as trade and development. Climate leadership therefore needs to be focused on creating synergies with other priorities and demonstrating the up-sides of climate protection. It will thereby improve the likelihood that industrialized countries that remain outside the multilateral climate regime and larger developing countries will take on robust climate commitments in the future. This paper identifies what a leadership coalition of countries might do to improve international willingness to address climate change through a set of recommendations under four priority areas for action."
- *China's National Climate Change Programme*
  People's Republic of China, 2007
  http://en.ndrc.gov.cn/newsrelease/P020070604561191006823.pdf
- *Climate Alarm: Disasters Increase as Climate Change Bites*
  Oxfam International, 2007
  http://oxfam.intelli-direct.com/e/d.dll?m=235&url=http://www.oxfam.org/en/files/bp108_climate_change_alarm_0711.pdf/download
  Summary: "Climatic disasters are increasing as temperatures climb and rainfall intensifies. A rise in small- and medium-scale disasters is a particularly worrying trend. Yet even extreme weather need not bring disasters; it is poverty and powerlessness that make people vulnerable. Though more emergency aid is needed, humanitarian response must do more than save lives: it has to link to climate change adaptation and bolster poor people's livelihoods through social protection and disaster risk reduction approaches."

- *Climate Change Activities in the United States*
  Pew Center on Global Climate Change, 2004
  http://www.pewclimate.org/docUploads/74241_US%20
  Activities%20Report_040604_075445.pdf
  Summary: "This report summarizes climate change efforts in the
  United States, including activity:
    - In Congress, where in October 2003, the U.S. Senate for the first
      time voted on legislation that would cap U.S. greenhouse gas
      (GHG) emissions and establish a national GHG trading system;
    - At the state level, where governments are enacting mandatory
      carbon controls and other programs to reduce emissions; and
    - In the business community, where a growing number of cor-
      porations are setting greenhouse gas targets and achieving
      significant emission reductions."
- *Climate Change and Disaster Management*
  Geoff O'Brien et. al., in *Disasters,* vol. 30, no. 1 (2006), 64–80
  http://www.blackwell-synergy.com/doi/pdf/10.1111/j.1467-
  9523.2006.00307.x (fee)
  Summary: "Climate change, although a natural phenomenon,
  is accelerated by human activities. Disaster policy response to
  climate change is dependent on a number of factors, such as
  readiness to accept the reality of climate change, institutions and
  capacity, as well as willingness to embed climate change risk
  assessment and management in development strategies. These
  conditions do not yet exist universally. A focus that neglects to
  enhance capacity-building and resilience as a prerequisite for
  managing climate change risks will, in all likelihood, do little
  to reduce vulnerability to those risks. Reducing vulnerability is
  a key aspect of reducing climate change risk. To do so requires
  a new approach to climate change risk and a change in institutional structures and relationships. A focus on development that
  neglects to enhance governance and resilience as a prerequisite
  for managing climate change risks will, in all likelihood, do little
  to reduce vulnerability to those risks."
- *Climate Change and Variability in California*
  National Center for Ecological Analysis and Synthesis, 1998
  http://www.nceas.ucsb.edu/ca/climate.pdf
  Excerpt: "The writing of this White Paper has provided a wonder-
  ful opportunity to develop a snapshot of California and to consider the potential impacts of climate change and variability for

the state. When we set forth on this project, I assumed that there was a basic state document that summarized the state's economy, its major sectors, and its basic features. Perhaps it exists, but it has eluded the efforts of the authors and advisors to discover it. In fact, it is surprisingly difficult to reconcile the many different figures given for key economic sectors and activities. Similarly, the precise condition of natural systems and the history of events is cloudy. The following pages seek to outline California's key economic sectors, important physical features, environmental conditions, and diverse population. It is based on official state sources, such as the California Trade and Commerce Agency's web site, and on numerous other published and electronic sources which are referenced and listed at the end of this document. The purpose of this summary is to provide a basis for consideration of potential impacts of climate change and variability on California. The focus is therefore on California-specific information. The broader issue of global climate change has been extensively documented in the literature. Rather than restate that information here, the reader is referred to both the official published sources, such as the Intergovernmental Panel on Climate Change (IPCC) documents, the scientific literature, and the numerous excellent web sites which are continuously updating information of the science and policy of climate change."

- *Climate Change Impacts on the United States: The Potential Consequences of Climate Change Variability and Change*
  National Assessment Synthesis Team, 2001
  http://www.usgcrp.gov/usgcrp/Library/nationalassessment/foundation.htm
  Excerpt: "The National Assessment of the Potential Consequences of Climate Variability and Change is a landmark in the major ongoing effort to understand what climate change means for the United States. Climate science is developing rapidly and scientists are increasingly able to project some changes at the regional scale, identifying regional vulnerabilities, and assessing potential regional impacts. Science increasingly indicates that the Earth's climate has changed in the past and continues to change, and that even greater climate change is very likely in the 21st century. This Assessment has begun a national process of research, analysis, and dialogue about the coming changes in climate, their impacts, and what Americans can do to adapt to an uncertain and continuously

changing climate. This Assessment is built on a solid foundation of science conducted as part of the United States Global Change Research Program (USGCRP). This document is the Foundation report, which provides the scientific underpinnings for the Assessment. It has been prepared in cooperation with independent regional and sector assessment teams under the leadership of the National Assessment Synthesis Team (NAST). The NAST is a committee of experts drawn from governments, universities, industry, and non-governmental organizations. It has been responsible for preparing an Overview report aimed at general audiences and for broad oversight of the Assessment along with the Federal agencies of the USGCRP. These two national-level, peer-reviewed documents synthesize results from studies conducted by regional and sector teams, and from the broader scientific literature."

- *Climate Change Futures: Health, Ecological, and Economic Dimensions*
  Harvard Medical School, 2005
  http://www.climatechangefutures.org/pdf/CCF_Report_Final_10.27.pdf
  Excerpt: "Climate is the context for life on earth. Global climate change and the ripples of that change will affect every aspect of life, from municipal budgets for snowplowing to the spread of disease. Climate is already changing, and quite rapidly. With rare unanimity, the scientific community warns of more abrupt and greater change in the future. Many in the business community have begun to understand the risks that lie ahead. Insurers and reinsurers find themselves on the front lines of this challenge since the very viability of their industry rests on the proper appreciation of risk. In the case of climate, however, the bewildering complexity of the changes and feedbacks set in motion by a changing climate defy a narrow focus on sectors. For example, the effects of hurricanes can extend far beyond coastal properties to the heartland through their impact on offshore drilling and oil prices. Imagining the cascade of effects of climate change calls for a new approach to assessing risk. The worst-case scenarios would portray events so disruptive to human enterprise as to be meaningless if viewed in simple economic terms. On the other hand, some scenarios are far more positive (depending on how society reacts to the threat of change). In addition to examining

current trends in events and costs, and exploring case studies of some of the crucial health problems facing society and the natural systems around us, 'Climate Change Futures: Health, Ecological and Economic Dimensions' uses scenarios to organize the vast, fluid possibilities of a planetary scale threat in a manner intended to be useful to policymakers, business leaders and individuals."

- *Climate Change: Adapt or Bust*
  Lloyds "360 Risk Project," 2006
  http://www.lloyds.com/NR/rdonlyres/38782611-5ED3-4FDC-85A4-5DEAA88A2DA0/0/FINAL360climatechangereport.pdf
  Summary: "Until recently, world opinion has been divided: are current weather trends the result of long-term climate change or not? And what role, if any, has climate change played in the recent spate of weather-related catastrophes? The facts are often confused by politics and by a wealth of different — and sometimes conflicting — evidence from a range of scientific and other sources. However, a growing body of expert opinion now agrees that the climate is changing — and that human activity is playing a major role. Most worryingly, the latest science suggests that future climate change may take place quicker than previously anticipated. There will continue to be much argument, over both the extent of future climate change and its likely impact on society. But whatever the facts, there could hardly be a debate of greater importance to the insurance industry."

- *Climate Change: Financial Risks to Federal and Private Insurers in Coming Decades are Potentially Significant*
  US Government Accountability Office, 2007
  http://www.gao.gov/new.items/d07285.pdf
  Summary: "Key scientific assessments report that the effects of climate change on weather-related events and, subsequently, insured and uninsured losses, could be significant. The key assessments GAO reviewed generally found that rising temperatures are expected to increase the frequency and severity of damaging weather-related events, such as flooding or drought, although the timing and magnitude are as yet undetermined. Taken together, private and federal insurers paid more than $320 billion in claims on weather-related losses from 1980 to 2005. Claims varied significantly from year to year — largely due to the effects of catastrophic weather events such as hurricanes and droughts — but have generally increased during this period. The

growth in population in hazard-prone areas and resulting real estate development have generally increased liabilities for insurers, and have helped to explain the increase in losses. Due to these and other factors, federal insurers' exposure has grown substantially. Since 1980, NFIP's exposure quadrupled, nearing $1 trillion in 2005, and program expansion increased FCIC's exposure 26-fold to $44 billion. Major private and federal insurers are both exposed to the effects of climate change over coming decades, but are responding differently. Many large private insurers are incorporating climate change into their annual risk management practices, and some are addressing it strategically by assessing its potential long-term industry-wide impacts. The two major federal insurance programs, however, have done little to develop comparable information. GAO acknowledges that the federal insurance programs are not profit-oriented, like private insurers. Nonetheless, a strategic analysis of the potential implications of climate change for the major federal insurance programs would help the Congress manage an emerging high-risk area with significant implications for the nation's growing fiscal imbalance."

- *Climate of Disaster*
  Tearfund, 2007
  http://www.tearfund.org/webdocs/Website/Campaigning/
  Policy%20and%20research/Climate%20of%20Disaster.pdf
  Excerpt: "The equivalent of a third of the world's population has already been affected by weather-related disasters and this is set to soar unless urgent international action is taken, including at least £25 billion spent every year helping the world's most vulnerable communities prepare to save their own lives. As Bangladesh reels from the recent cyclone in which millions of people remain affected, it is clearer than ever before that the world must change the way it tackles weather-related disasters or face catastrophic consequences. Airlifting stranded people from floodwaters and sending food packages to those affected by drought can no longer be our sole response to weather-related disasters. As a global community we have a moral responsibility to invest our aid money upfront in helping the planet's poorest people prepare for disaster. If we do not, then many thousands of lives will be needlessly lost and billions of pounds of aid money will not be used to best effect. Climate change is already increasing the number and intensity of extreme events such as floods and droughts. This has

resulted in more disasters affecting millions of the world's most vulnerable people."

- *Confronting Climate Change: Avoiding the Unmanageable and Managing the Unavoidable*
  United Nations Foundation, 2007
  http://www.unfoundation.org/files/pdf/2007/SEG_Report.pdf
  Summary: "Global climate change, driven largely by the combustion of fossil fuels and by deforestation, is a growing threat to human well-being in developing and industrialized nations alike. Significant harm from climate change is already occurring, and further damages are a certainty. The challenge now is to keep climate change from becoming a catastrophe. There is still a good chance of succeeding in this, and of doing so by means that create economic opportunities that are greater than the costs and that advance rather than impede other societal goals. But seizing this chance requires an immediate and major acceleration of efforts on two fronts: mitigation measures (such as reductions in emissions of greenhouse gases and black soot) to prevent the degree of climate change from becoming unmanageable; and adaptation measures (such as building dikes and adjusting agricultural practices) to reduce the harm from climate change that proves unavoidable."

- *Disaster Risk, Climate Change and International Development: Scope For, and Challenges to, Integration*
  Lisa Schipper and Mark Pelling, in *Disasters*, vol. 30, no. 1 (2006), 1–4
  http://www.blackwell-synergy.com/doi/pdf/10.1111/j.1467-9523.2006.00304.x
  Summary: "Reducing losses to weather-related disasters, meeting the Millennium Development Goals and wider human development objectives, and implementing a successful response to climate change are aims that can only be accomplished if they are undertaken in an integrated manner. Currently, policy responses to address each of these independently may be redundant or, at worst, conflicting. We believe that this conflict can be attributed primarily to a lack of interaction and institutional overlap among the three communities of practice. Differences in language, method and political relevance may also contribute to the intellectual divide. Thus, this paper seeks to review the theoretical and policy linkages among disaster risk reduction, climate change

and development. It finds that not only does action within one realm affect capacity for action in the others, but also that there is much that can be learnt and shared between realms in order to ensure a move towards a path of integrated and more sustainable development."

- *Fighting Climate Change: Human Solidarity in a Divided World*
  United Nations Development Programme, 2007
  http://hdr.undp.org/en/media/hdr_20072008_summary_english.
  pdf
  Excerpt: "What we do today about climate change has consequences that will last a century or more. The part of that change that is due to greenhouse gas emissions is not reversible in the foreseeable future. The heat trapping gases we send into the atmosphere in 2008 will stay there until 2108 and beyond. We are therefore making choices today that will affect our own lives, but even more so the lives of our children and grandchildren. This makes climate change different and more difficult than other policy challenges. Climate change is now a scientifically established fact. The exact impact of greenhouse gas emission is not easy to forecast and there is a lot of uncertainty in the science when it comes to predictive capability. But we now know enough to recognize that there are large risks, potentially catastrophic ones, including the melting of ice-sheets on Greenland and the West Antarctic (which would place many countries under water) and changes in the course of the Gulf Stream that would bring about drastic climatic changes."

- *Health Effects of Climate Change in the UK*
  Expert Group on Climate Change and Health in the UK, 2001
  http://www.dh.gov.uk/prod_consum_dh/idcplg?IdcService=
  GET_FILE&dID=1733&Rendition=Web
  Summary: "At the request of the DH, the Expert Group on Climate Change and Health in the UK reported on the likely impact of climate change on health, and implications for the NHS. The report discusses public perceptions of the impact of climate change on health, and available methods for assessing health implications of climate change. It goes on to present an overview of the subject, and to discuss potential effects of measures aimed at mitigating climate change. It makes a series of tentative predictions relating to cold and heat-related deaths, food poisoning, vector-borne and water-borne diseases, disasters caused by gales and coastal

flooding, effects of air pollutants and ozone, skin cancer, and measures to reduce greenhouse gas emissions. An annex lists members of the Expert Group. References cited at the end of each chapter."

- *Impacts of Climate Change*
  Nils GIlman, Doug Randall, and Peter Schwartz (Global Business Network, 2007)
  http://www.gbn.com/climatechange/ImpactsOfClimateChange.pdf
  Summary: "In this paper we explore several of the possible impacts of continued, relatively unrestrained greenhouse gas emissions over the next half-century. These impacts, although not always highly likely, are plausible. In particular, we focus on already stressed systems that are vulnerable to being driven over the edge or past a tipping point by either radical or gradual shifts in climate. By doing so, we offer an alternative analytic approach — a 'system vulnerability' approach — to understanding and anticipating climate change disruptions. We conclude by considering both the security implications of the climate impacts discussed in this paper, and the analytic opportunities provide by the systems vulnerability approach."

- *Livelihoods and Climate Change: Combining Disaster Risk Reduction, Natural Resource Management, and Climate Change Adaptation in a New Approach to the Reduction of Vulnerability and Poverty*
  Task Force on Climate Change, Vulnerable Communities and Adaptation, 2003
  http://www.iisd.org/pdf/2003/natres_livelihoods_cc.pdf
  Summary: "Whatever happens to future greenhouse gas emissions, we are now locked into inevitable changes to climate patterns. Adaptation to climate change is therefore no longer a secondary and long-term response option only to be used as a last resort. It is now prevalent and imperative, and for those communities already vulnerable to the impacts of present day climate hazards, an urgent imperative."

- *Local Initiatives and Adaptation to Climate Change*
  Ana V. Rojas Blanco, in *Disasters, vol. 30, no. 1 (2006), 140–47*
  http://www.blackwell-synergy.com/doi/pdf/10.1111/j.1467-9523.2006.00311.x
  Summary: "Climate change is expected to lead to an increase in the number and strength of natural hazards produced by climatic

events. This paper presents some examples of the experiences of community-based organisations (CBOs) and non-governmental organisations (NGOs) of variations in climate, and looks at how they have incorporated their findings into the design and implementation of local adaptation strategies. Local organisations integrate climate change and climatic hazards into the design and development of their projects as a means of adapting to their new climatic situation. Projects designed to boost the resilience of local livelihoods are good examples of local adaptation strategies. To upscale these adaptation initiatives, there is a need to improve information exchange between CBOs, NGOs and academia. Moreover, there is a need to bridge the gap between scientific and local knowledge in order to create projects capable of withstanding stronger natural hazards."

- *Meeting the Climate Challenge: Recommendations of the International Climate Change Task Force*
  The International Climate Change Task Force, 2005
  http://www.americanprogress.org/kf/climatechallenge.pdf
  Summary: "To chart a way forward, an International Climate Change Taskforce, composed of leading scientists, public officials, and representatives of business and non-governmental organizations, was established at the invitation of three leading public policy institutes — the Institute for Public Policy Research, the Center for American Progress and The Australia Institute. The Taskforce's aim has been to develop proposals to consolidate and build on the gains achieved under the UNFCCC and the Kyoto Protocol to ensure that climate change is addressed effectively over the long term. In doing so, the Taskforce has met twice, in Windsor, United Kingdom and Sydney, Australia, where we reviewed and debated detailed research papers prepared by the Taskforce Secretariat, provided by the three founding organizations. The Taskforce's recommendations are to all governments and policy-makers worldwide. However, particular emphasis is placed on providing independent advice to the governments of the Group of Eight (G8) and the European Union (EU) in the context of the UK's presidencies of both organizations in 2005, during which Prime Minister Tony Blair has pledged to make addressing climate change a priority. The recommendations are also made in the context of the start of international negotiations in 2005 on future collective action on climate change, and the need to engage

239

the governments of those industrialized countries that have not ratified the Kyoto Protocol. The Taskforce's recommendations are presented in the report."

- *Natural Disasters and Climate Change*
  Madeleen Helmer and Dorothea Hilhorst, in *Disasters*, vol. 30, no. 1 (2006), 1–4
  http://www.blackwell-synergy.com/doi/pdf/10.1111/j.1467-9523.2006.00302.x
  Summary: "Human emissions of greenhouse gases are already changing our climate. This paper provides an overview of the relation between climate change and weather extremes, and examines three specific cases where recent acute events have stimulated debate on the potential role of climate change: the European heatwave of 2003; the risk of inland flooding, such as recently in Central Europe and Great Britain; and the harsh Atlantic hurricane seasons of 2004 and 2005. Furthermore, it briefly assesses the relation between climate change and El Niño, and the potential of abrupt climate change. Several trends in weather extremes are sufficiently clear to inform risk reduction efforts. In many instances, however, the potential increases in extreme events due to climate change come on top of alarming rises in vulnerability. Hence, the additional risks due to climate change should not be analysed or treated in isolation, but instead integrated into broader efforts to reduce the risk of natural disasters."

- *No Place to Hide: Effects of Climate Change on Protected Areas*
  World Wildlife Foundation Climate Change Programme, 2003
  http://assets.panda.org/downloads/wwfparksbro.pdf
  Excerpt: "Protected area agencies could be faced with the massive task of having to shift protected areas to keep up with moving habitats and ecosystems. Some protected areas may have to retrench onto higher ground as water rises. The practical difficulties should not be underestimated. Protected areas do not exist in an empty landscape and replacement land and water will often not be available. 'Moving' protected areas would have enormous implications for their infrastructure, surrounding human communities and the many businesses associated with parks. Shifting reserves would have cultural implications; societies build powerful emotional bonds to national parks and nature reserves that mean they cannot simply be swapped and replaced lightly. We are still learning about climate change and there have been relatively few

studies of impacts within protected areas to confirm or disprove the modeling exercises and speculation. Ecosystems are often quite resilient but while some climate change problems are likely to be surmountable through management, adaptation or evolution, others are likely to be more intractable."

- *Observed Variability and Trends in Extreme Climate Events: A Brief Review*
  Bulletin on the American Meteorological Society, 2000
  http://www.ncdc.noaa.gov/oa/pub/data/special/extr-bams2.pdf
  Summary: "Variations and trends in extreme climate events have only recently received much attention. Exponentially increasing economic losses, coupled with an increase in deaths due to these events, have focused attention on the possibility that these events are increasing in frequency. One of the major problems in examining the climate record for changes in extremes is a lack of high quality long-term data. In some areas of the world increases in extreme events are apparent, while in others there appears to be a decline. Based on this information increased ability to monitor and detect multi-decadal variations and trends is critical to begin to detect any observed changes and understand their origins."

- *Oxfam Analysis of the Bali Conference Outcomes*
  Oxfam International, 2007
  http://www.oxfam.org.uk/resources/policy/climate_change/downloads/bali_analysis.pdf

- *Pathways to Energy & Climate Change 2050*
  World Business Council for Sustainable Development, 2006
  http://www.wbcsd.org/web/publications/pathways.pdf
  Excerpt: "Pathways to 2050: Energy and Climate Change builds on the WBCSD's 2004 Facts and Trends to 2050: Energy and Climate Change and provides a more detailed overview of potential pathways to reducing CO_2 emissions. The pathways shown illustrate the scale and complexity of the change needed, as well as the progress that has to be made through to 2050. Our 'checkpoint' in 2025 gives a measure of this progress and demonstrates the urgency to act early to shift to a sustainable emissions trajectory. The WBCSD has chosen to continue to illustrate the challenges associated with one particular trajectory, consistent with the discussion already presented in Facts and Trends. This document therefore looks closely at the changes needed to begin to stabilize CO_2 concentrations in the atmosphere at no more than 550-ppm

(see glossary), which relates to the '9 Gt world' described in Facts and Trends. As such, and based upon simplified assumptions and extrapolations, we have made many choices, some arbitrary, to present this single illustrative story. It is neither a fully-fledged scenario nor does it recommend a target. Moreover, this document does not discuss policy definitions or options, topics that need to be dealt with separately."

- *State of the Climate*
  National Climatic Data Center, US Dept. of Commerce, Annual Report
  http://www.ncdc.noaa.gov/oa/climate/research/monitoring.html#state
- *Stern Review on the Economics of Climate Change*
  Treasury, Government of the United Kingdom, 2007
  http://www.hm-treasury.gov.uk/independent_reviews/stern_review_economics_climate_change/stern_review_report.cfm
  Excerpt: "There is still time to avoid the worst impacts of climate change, if we take strong action now. The scientific evidence is now overwhelming: climate change is a serious global threat, and it demands an urgent global response. This Review has assessed a wide range of evidence on the impacts of climate change and on the economic costs, and has used a number of different techniques to assess costs and risks. From all of these perspectives, the evidence gathered by the Review leads to a simple conclusion: the benefits of strong and early action far outweigh the economic costs of not acting. Climate change will affect the basic elements of life for people around the world — access to water, food production, health, and the environment. Hundreds of millions of people could suffer hunger, water shortages and coastal flooding as the world warms. Using the results from formal economic models, the Review estimates that if we don't act, the overall costs and risks of climate change will be equivalent to losing at least 5% of global GDP each year, now and forever. If a wider range of risks and impacts is taken into account, the estimates of damage could rise to 20% of GDP or more. In contrast, the costs of action — reducing greenhouse gas emissions to avoid the worst impacts of climate change — can be limited to around 1% of global GDP each year. The investment that takes place in the next 10–20 years will have a profound effect on the climate in the second half of this century and in the next. Our actions now and over the coming decades could

create risks of major disruption to economic and social activity, on a scale similar to those associated with the great wars and the economic depression of the first half of the 20th century. And it will be difficult or impossible to reverse these changes. So prompt and strong action is clearly warranted. Because climate change is a global problem, the response to it must be international. It must be based on a shared vision of long-term goals and agreement on frameworks that will accelerate action over the next decade, and it must build on mutually reinforcing approaches at national, regional and international level."

- *The Impacts of Climate Change on the Risk of Natural Disasters*
  Martin K. Aalst, in *Disasters*, vol. 30, no. 1 (2006), 5–18
  http://www.blackwell-synergy.com/doi/pdf/10.1111/j.1467-9523.2006.00303.x
  Summary: "Human emissions of greenhouse gases are already changing our climate. This paper provides an overview of the relation between climate change and weather extremes, and examines three specific cases where recent acute events have stimulated debate on the potential role of climate change: the European heatwave of 2003; the risk of inland flooding, such as recently in Central Europe and Great Britain; and the harsh Atlantic hurricane seasons of 2004 and 2005. Furthermore, it briefly assesses the relation between climate change and El Niño, and the potential of abrupt climate change. Several trends in weather extremes are sufficiently clear to inform risk reduction efforts. In many instances, however, the potential increases in extreme events due to climate change come on top of alarming rises in vulnerability. Hence, the additional risks due to climate change should not be analysed or treated in isolation, but instead integrated into broader efforts to reduce the risk of natural disasters."

- *United Nations Environmental Programme (UNEP) Intergovernmental Panel on Climate Change (IPCC) Working Group Report "Mitigation of Climate Change"*
  United Nations Environmental Programme (UNEP), 2007
  http://www.ipcc.ch/ipccreports/ar4-wg3.htm
  Summary: "The Working Group III contribution to the IPCC Fourth Assessment Report (AR4) focuses on new literature on the scientific, technological, environmental, economic and social aspects of mitigation of climate change, published since the IPCC Third Assessment Report (TAR) and the Special Reports on CO_2

Capture and Storage (SRCCS) and on Safeguarding the Ozone Layer and the Global Climate System (SROC). The report is organized into six sections:
– Mitigation in the long-term (beyond 2030)
– Mitigation in the short and medium term, across different economic sectors (until 2030)
– Policies, measures and instruments to mitigate climate change
– Sustainable development and climate change mitigation
– Greenhouse gas (GHG) emission trends
– Gaps in knowledge."
• *Unnatural Disaster: Global Warming and Our National Parks* **National Parks Conservation Association, 2007**
http://www.npca.org/globalwarming/unnatural_disaster.pdf
Summary: "Although the situation seems dire, we can still halt the most severe effects of global warming if we take action now. The Centennial anniversary of the National Park System in 2016 provides sufficient time and a symbolically important deadline in which to act. Federal, state, and local governments, along with individuals, can take actions within that timeframe that will slow and in some cases halt the damage. Over the next nine years, the national parks offer a unique opportunity to draw attention to America's priceless resources at risk, and to showcase opportunities to act to protect them. As chronicled in this report, national parks already are helping us to understand how global warming affects our natural world. Within them, we see the warning signs of major changes ahead. We must learn how to manage parks to maintain healthy ecosystems in the face of climate change, and we must build public support for doing so."
• *White Paper on the Ethical Dimensions of Climate Change* **Rock Ethics Institute, Penn State University**
http://rockethics.psu.edu/climate/whitepaper/edcc-whitepaper.pdf
Excerpt: "This paper describes the relevant facts, ethical questions, and preliminary ethical analyses that will constitute the initial phase of the Collaborative Program on the Ethical Dimensions of Climate (EDCC). This paper does not seek to deal with these matters exhaustively but rather intends to create a focus for initial inquiry and draw preliminary conclusions about the ethical dimensions of several climate change issues that are possible at this early stage of the work of the EDCC. By the use of the word 'ethics' in this paper is meant the field of philosophical inquiry

that examines concepts and their employment about what is right and wrong, obligatory and non-obligatory, and when responsibility should attach to human actions that cause harm. For this reason, an ethical examination of climate change issues will explore prescriptive assertions about what should be done about climate change rather than focus on descriptions of scientific and economic facts alone, although good ethical analyses of climate change issues must be sensitive to facts that frame any issue. For this reason, this paper identifies the scientific, economic, and social facts associated with each issue about which it draws ethical conclusions."

WEB SITES

- Alaska Climate Change Web site
  http://esp.cr.usgs.gov/info/assessment/alaska.html
- Argentina's National Climate Change Web site
  http://www.medioambiente.gov.ar/default.asp?idseccion=29
- Arizona Climate Change Web site
  http://www.azclimatechange.us/background-impacts.cfm
- Asia-Pacific Network for Global Change Research (APN)
  http://www.apn-gcr.org/en/indexe.html
- Atlantic Climate Change Program
  http://www.aoml.noaa.gov/phod/accp/
- Atmosphere, Climate & Environment Information Programme of the UK Department for Environment, Food and Rural Affairs (DEFRA)
  http://www.ace.mmu.ac.uk/
- Australia National Climate Change Web site
  http://www.dfat.gov.au/environment/climate/
- Austrian Council on Climate Change (ACCC)
  http://www.accc.gv.at/
- Bahamas National Climate Change Web site
  http://www.best.bs/index.html
- Berkeley Lab Earth Sciences Division Climate Change Program
  http://www-esd.lbl.gov/CLIMATE/index.html
- Brazil National Climate Change Web site
  http://www.mct.gov.br/index.php/content/view/3881.html

- Bulgaria Climate Change Web site
  http://www.climatechange.be/
- California Climate Change Portal
  http://www.climatechange.ca.gov/
- Cambodia Office of Climate Change
  http://www.camclimate.org.kh/
- Canada National Climate Change Web site
  http://climatechange.gc.ca/
- Carbon Dioxide Information Analysis Center
  http://cdiac.esd.ornl.gov/
- Center for Global Environmental Research
  http://www-cger.nies.go.jp/
- Chile National Climate Change Web site
  http://www.conama.cl/portal/1301/channel.html
- China National Climate Change Program
  http://www.10thnpc.org.cn/english/environment/213624.htm
- Cities for Climate Protection Initiatives
  http://www.ci.duluth.mn.us/city/information/ccp/index.htm
- City of Ann Arbor (Michigan) Climate Protection Initiatives
  http://www.cambridgema.gov/~CDD/et/env/climate/climate.
  html#plan
- City of Regina (Canada) Climate Change Program
  http://www.regina.ca/content/info_services/climate/information/
  index.shtml
- Clean Cities Program (US Department of Energy)
  http://www.ccities.doe.gov/
- Climate Change and Health Web site
  http://www.hc-sc.gc.ca/ewh-semt/pubs/climat/index_e.html
- Climate Change Chronicles
  http://www.climatechange.com.au/
- Climate Change Prediction Program
  http://www.csm.ornl.gov/chammp/
- Climate Change Technology Program
  http://www.climatetechnology.gov/
- Climate Institute
  http://www.climate.org/climate_main.shtml
- Climate VISION ("Voluntary Innovative Sector Initiatives: Opportunities Now")
  http://www.climatevision.gov/

- Colombia National Climate Change Web site
  http://www.minambiente.gov.co/
- Colorado Climate Change Web site
  http://www.cdphe.state.co.us/ap/planning.asp#Colorado%20
  climate%20change
- Connecticut Climate Change Web site
  http://www.ctclimatechange.com/
- Convention on Biological Diversity
  http://www.biodiv.org/
- Czech Republic Climate Change Web site
  http://www.ctclimatechange.com/
- Denmark Climate Change Web site
  http://www.mst.dk/forside/
- El Salvador National Climate Change Web site
  http://www.marn.gob.sv/cambio_climatico.htm
- Encyclopedia of the Atmospheric Environment
  http://www.ace.mmu.ac.uk/eae/english.html
- European Climate Change Programme
  http://ec.europa.eu/environment/climat/eccp.htm
- Florida Division of Air Resource Management: Greenhouse Effect
  and Global Warming Web site
  http://www.dep.state.fl.us/air/pollutants/greenhouse.htm
- Former Yugoslav Republic of Macedonia National Climate Change
  Web site
  http://www.unfccc.org.mk/
- France National Climate Change Web site
  http://www.ecologie.gouv.fr/sommaire.php3
- Georgia National Climate Change Web site
  http://www.ccna.caucasus.net/
- Germany Climate Change Web site
  http://www.bmu.de/english/aktuell/4152.php
- Global Change Data Center
  http://tsdis02.nascom.nasa.gov/gcdc/
- Global Change Data Information Systems
  http://globalchange.gov/
- Global Change Master Directory
  http://gcmd.gsfc.nasa.gov/
- Global Earth Observatory System of Systems
  http://www.epa.gov/geoss/

- Global Environment Facility
  http://www.gefweb.org/
- Government of Canada's Climate Change Impacts and Adaptation Program
  http://adaptation.nrcan.gc.ca/home_e.asp
- Greece National Climate Change Web site
  http://www.minenv.gr/4/41/e4100.html
- Guide to Geologic Change in Alaska: Weather Fluctuations and Climate Change
  http://www.dggs.dnr.state.ak.us/geologic_hazards_climate.htm
- Haiti National Climate Change Web site
  http://unfccc.int/resource/ccsites/haiti/ccweb/index.html
- Harvard Medical School Center for Health and the Global Environment: Climate Change Program
  http://chge.med.harvard.edu/programs/ccf/index.html
- Hawaii Climate Change Web site
  http://www.hawaii.gov/dbedt/czm/wec/html/weather/climate.htm
- Hungary National Climate Change Web site
  http://www.ktm.hu/
- Illinois Global Climate Change Project
  http://dnr.state.il.us/orep/inrin/eq/iccp/iccp.htm
- India National Climate Change Web site
  http://envfor.nic.in/cc/index.htm
- Institute for the Study of Society and Environment
  http://www.isse.ucar.edu/index.jsp
- Inter-American Institute for Global Change Research
  http://www.iai.int/
- Intergovernmental Panel on Climate Change, WMO
  http://www.ipcc.ch/
- International Geosphere — Biosphere Programme
  http://www.igbp.kva.se/cgi-bin/php/frameset.php
- Ireland National Climate Change Web site
  http://www.environ.ie/DOEI/DOEIPol.nsf/wvNavView/Climate+Change?OpenDocument&Lang=
- Jordan and the UN Framework Convention on Climate Change
  http://unfccc.int/resource/ccsites/jordan/index.html
- Latvia National Climate Change Web site
  http://www.varam.gov.lv/

- Lebanon National Climate Change Web site
  http://www.moe.gov.lb/ClimateChange/index.html
- Linking Climate Adaptation Network
  http://www.linkingclimateadaptation.org/
- Lithuania National Climate Change Web site
  http://www.am.lt/VI/
- Luxemburg National Climate Change Web site
  http://www.environnement.public.lu/
- Maine Climate Change Web site
  http://www.maine.gov/dep/air/globalwarming/
- Marshall Islands Climate Change Web site
  http://unfccc.int/resource/ccsites/marshall/index.html
- Mexico National Climate Change Web site
  http://www.semarnat.gob.mx/spp/sppa/dgapcc/c_index.htm
- Ministry of Nature Protection of the Republic of Armenia's Climate
  Change Information Center
  http://www.nature-ic.am/
- Minnesota Climate Change Web site
  http://www.moea.state.mn.us/reduce/climatechange.cfm
- Missouri Climate Change Web site
  http://www.dnr.mo.gov/energy/cc/cc.htm
- Morocco Climate Change Web site
  http://www.ccmaroc.ma/
- NASA Earth Observatory
  http://earthobservatory.nasa.gov/
- NASA Global Change Master Directory Learning Center
  http://gcmd.nasa.gov/Resources/Learning/
- National Academy of Sciences Board on Atmospheric Sciences
  and Climate
  http://dels.nas.edu/basc/
- National Environmental Research Council Rapid Climate Change
  Web site
  http://www.noc.soton.ac.uk/rapid/rapid.php
- New Hampshire Department of Environmental Services Climate
  Change Web site
  http://www.des.state.nh.us/ard/climatechange/
- New Jersey Climate Change Web site
  http://www.state.nj.us/dep/dsr/climate/climate.htm
- New Mexico Climate Change Web site
  http://www.nmclimatechange.us/background-impacts.cfm

- New Zealand National Climate Change Web site
  http://www.mfe.govt.nz/issues/climate/
- NOAA National Climatic Data Center
  http://www.ncdc.noaa.gov/
- Northern Ireland National Climate Change Web site
  http://www.ehsni.gov.uk/environment/climatechange/air-climate-change.shtml
- Norway National Climate Change Web site
  http://www.environment.no/templates/PageWithRightListing.aspx?id=2142
- Official Web site of Al Gore's Documentary *An Inconvenient Truth*
  http://www.climatecrisis.net/
- Oregon Climate Change Web site
  http://www.deq.state.or.us/
- Oregon Climate Change Web site
  http://www.oregon.gov/ENERGY/GBLWRM/climhme.shtml
- Oxfam International Climate Change Site
  http://www.oxfam.org.uk/resources/issues/climatechange/introduction.html
- Poland Climate Change Web site
  http://www.mos.gov.pl/
- Reducing Canada's Vulnerability to Climate Change
  http://ess.nrcan.gc.ca/2002_2006/rcvcc/index_e.php
- Regional Environmental Center for Central and Eastern Europe Climate Change Programme
  http://www.rec.org/REC/Programs/ClimateChange.html
- Romania Climate Change Web site
  http://www.mmediu.ro/dep_mediu/schimbari_climatice/schimbari_climatice.htm
- Science Policy Assessment and Research on Climate (University of Colorado)
  http://sciencepolicy.colorado.edu/sparc/
- Senegal "UN Framework Convention on Climate Change" Web site
  http://unfccc.int/resource/ccsites/senegal/index.htm
- Seychelles "UN Framework Convention on Climate Change" Web site
  http://unfccc.int/resource/ccsites/seychell/
- Singapore National Climate Change Web site
  http://app.mewr.gov.sg/home.asp?id=M1
- Slovakia National Climate Change Web site
  http://www.enviro.gov.sk/servlets/page/166

- South Africa National Climate Change Web site
  http://www.environment.gov.za/ClimateChange2005/home.htm
- Spain National Climate Change Web site
  http://www.mma.es/portal/secciones/cambio_climatico/
- Sweden National Climate Change Web site
  http://www.internat.naturvardsverket.se/
- Switzerland National Climate Change Web site
  http://www.umwelt-schweiz.ch/buwal/eng/fachgebiete/klima/
  index.html
- The Royal Society Climate Change Web site
  http://royalsociety.org/landing.asp?id=1278
- UN Gateway to Climate Change
  http://www.un.org/climatechange/index.shtml
- Union of Concerned Scientists Gulf Coast Climate Change Web site
  http://www.ucsusa.org/gulf
- United Kingdom Climate Impacts Programme
  http://www.ukcip.org.uk/default.asp
- United Kingdom National Climate Change Web site
  http://www.defra.gov.uk/environment/climatechange/index.htm
- United Nations Environmental Programme Climate Change Web site
  http://www.unep.org/themes/climatechange/
- United Nations Framework Convention on Climate Change
  http://unfccc.int/2860.php
- University of Colorado "Societal Aspects of Weather" Web site
  http://sciencepolicy.colorado.edu/socasp/
- University of Washington Program on Climate Change
  http://www.uwpcc.washington.edu/
- Uruguay National Climate Change Web site
  http://www.cambioclimatico.gub.uy/index.php
- U.S. Agency for International Development Climate Change Program
  http://www.usaid.gov/our_work/environment/climate/
- U.S. Climate Change Science Program
  http://www.climatescience.gov/
- U.S. Climate Change Technology Program
  http://www.climatetechnology.gov/
- U.S. Department of Agriculture Global Change Program
  http://www.usda.gov/oce/global_change/index.htm

- U.S. Department of Commerce National Climatic Data Center (NCDC) Climate Monitoring Page
  http://www.ncdc.noaa.gov/oa/climate/research/monitoring.html
- U.S. Department of State Climate Change Web site
  http://www.state.gov/g/oes/climate/
- U.S. Department of Transportation Center for Climate Change and Environmental Forecasting
  http://climate.volpe.dot.gov/
- U.S. Environmental Protection Agency Climate Change Site
  http://epa.gov/climatechange/index.html
- U.S. Environmental Protection Agency Student Center
  http://epa.gov/students/
- U.S. Environmental Protection Agency Teaching Center
  http://epa.gov/teachers/
- U.S. Forest Service Climate Change Tree Atlas
  http://www.fs.fed.us/ne/delaware/atlas/index.html
- U.S. Geological Survey Teacher Packs on Global Change
  http://erg.usgs.gov/isb/pubs/teachers-packets/globalchange/globalhtml/
- U.S. Global Change Research Program
  http://www.usgcrp.gov/
- U.S. National Oceanographic and Atmospheric Administration Climate Change Web site
  http://lwf.ncdc.noaa.gov/oa/climate/climateextremes.html
- U.S. National Park Service Climate Change Web site
  http://www.nature.nps.gov/criticalissues/globalclimatechange.cfm
- USA Today Climate Change Resources
  http://www.usatoday.com/weather/resources/climate/climate-sci-resources.htm
- Venezuela National Climate Change Web site
  http://unfccc.int/resource/ccsites/venezuel/
- Vermont Climate Change Web site
  http://www.anr.state.vt.us/air/Planning/htm/ClimateChange.htm
- Washington State Climate Change Web site
  http://www.ecy.wa.gov/programs/air/globalwarming/Global_Warming_site.html
- White House Climate Change Policy
  http://www.whitehouse.gov/ceq/global-change.html

- World Bank Climate Change Web site
  http://web.worldbank.org/WBSITE/EXTERNAL/TOPICS/
  ENVIRONMENT/EXTCC/0,,menuPK:407870~pagePK:149018~piPK:
  149093~theSitePK:407864,00.html
- World Climate Research Program CLIVAR Climate Variability
  Program
  http://www.clivar.org/
- World Health Organisation Climate and Health Web site
  http://www.who.int/topics/climate/en/
- World Wildlife Foundation Climate Change Program
  http://www.panda.org/epo/
- Zambia "UN Framework Convention on Climate Change" Web site
  http://unfccc.int/resource/ccsites/zambia/
- Zimbabwe "UN Framework Convention on Climate Change"
  Web site
  http://unfccc.int/resource/ccsites/zimbab/

OTHER MATERIALS

- Guide
  A Consumer's Guide to Energy Efficiency and Renewable Energy
  U.S. Department of Energy
  http://www.eere.energy.gov/consumer/
- Guide
  A Guide to Climate Change for Small- to Medium-Sized Businesses
  The Canadian Chamber of Commerce
  http://www.pollutionprobe.org/Reports/Guide%20to%20CC%20
  for%20SMEs.pdf
- Guide
  Beginners Guide to the UN Framework Convention on Climate Change
  http://unfccc.int/resource/beginner.html
- Guide
  *Climate Change — An Australian Guide to the Science and Potential
  Impacts*
  Government of Australia
  http://www.greenhouse.gov.au/science/guide/index.html
- Guide
  Climate Change Controversies: A Simple Guide
  The Royal Society

http://royalsociety.org/page.asp?id=6229
- Guide
Climate Change Impacts and Risk Management
Government of Australia
http://www.greenhouse.gov.au/impacts/publications/pubs/risk-management.pdf
- Guide
Communicating and Learning About Global Climate Change
American Association for the Advancement of Science
http://www.aaas.org/news/press_room/climate_change/mtg_200702/climate_change_guide_2061.pdf
- Guide
Global Climate Change Student Guide
Manchester Metropolitan University
http://www.ace.mmu.ac.uk/Resources/gcc/contents.html
- Guide
Preparing for Climate Change: A Guide for Local Government in New Zealand
New Zealand Ministry for the Environment
http://www.mfe.govt.nz/publications/climate/preparing-for-climate-change-jul04/preparing-for-climate-change.pdf
- Guide
Reporting on Climate Change: Understanding the Science
National Safety Council Environmental Health Center
http://www.nsc.org/EHC/guidebks/climtoc.htm
- Guide
Surviving Climate Change in Small Islands
Tyndall Centre for Climate Change Research, 2005
http://www.tyndall.ac.uk/publications/surviving.pdf
- Guide
Teachers' Guide to High Quality Educational Materials on Climate Change and Global Warming
National Science Teachers Association
http://hdgc.epp.cmu.edu/teachersguide/teachersguide.htm
- Guide
The First 10 Years: An Overview of Actions Taken during the Past Decade to Combat Climate Change and Mitigate its Adverse Effects
United Nations Framework Convention on Climate Change
http://unfccc.int/resource/docs/publications/first_ten_years_en.pdf

- Guide
  Uniting on Climate 2007
  United Nations Framework Convention on Climate Change
  http://unfccc.int/files/essential_background/background_
  publications_htmlpdf/application/pdf/pub_07_uniting_on_
  climate_en.pdf
- Guide
  Working 9 to 5 on Climate Change
  World Resources Institute
  http://archive.wri.org/publication_detail.cfm?pubid=3756#1
- Guide
  Your Guide to Communicating Climate Change
  UK Department for Environment, Food, and Rural Affairs
  http://education.staffordshire.gov.uk/NR/rdonlyres/90C73AD7-
  82C3-4D3C-89E6-748008FD90E1/43010/communicating_climate_
  change.pdf
- Handbook
  Climate Change Partnership Handbook
  Alliance of Religions and Conservation
  http://www.arcworld.org/downloads/ClimateChange.pdf
- Map
  Global Warming: Early Warning Signs Map
  Climate Hot Map
  http://www.climatehotmap.org/
- Policy Document
  Adapting to Climate Change in Europe — Options for EU Action
  European Economic and Social Committee and the Committee of
  the Regions, 2007
  http://eur-lex.europa.eu/LexUriServ/LexUriServ.do?uri=COM:
  2007:0354:FIN:EN:PDF
- Policy Document
  United Nations Framework Convention on Climate Change
  United Nations, 1992
  http://unfccc.int/resource/docs/convkp/conveng.pdf
- Video
  "The Most Terrifying Video You'll Ever See"
  Greg Craven, Independence, OR, science teacher
  http://www.youtube.com/watch?v=zORv8wwiadQ&eurl=
  http://www.greenlivingtips.com/blogs/179/Global-climate-
  change-video.html

- Video
  A Way Forward: Facing Climate Change
  National Geographic
  http://video.nationalgeographic.com/video/player/environ-ment/global-warming-environment/way-forward-climate.html
- Video
  Climate Change and the New Millennium Development Goals: Meeting the Development
  Challenge. The Earth Institute at Columbia University, November 2007
  Session 1: http://www.dkv.columbia.edu/video/ei/ei_climate-change_680_sess1.html
  Session 2: http://www.dkv.columbia.edu/video/ei/ei_climate-change_680_sess2.html
- Video
  Global Climate Change and Human Well-Being
  American Association for the Advancement of Science
  RealVideo:http://www.aaas.org/news/press_room/climate_change/mtg_200702/climate_change_2007a.ram
  WindowsMedia:http://www.aaas.org/news/press_room/climate_change/media/climate_change_2007a.wmv
- Video
  Living With Climate Change
  European Union
  http://www.tvlink.org/viewer.cfm?vidID=251
- Video
  Practical Approaches to Climate Change
  Jeffrey D. Sachs, The Earth Institute at Columbia University, November 2006
  http://www.earth.columbia.edu/sitefiles/Media/events/video_archive/darfur.ram
- Video
  Understanding Climate Change
  Dr. Lonnie Thompson, School of Earth Sciences, Ohio State University
  Real Video: http://www.aaas.org/news/press_room/climate_change/mtg_200702/thompson.ram

INDEX

Intergovernmental Panel on
    Climate Change (IPCC) best
    case scenario, 34
need for sensible government
    policies, 48
planning approaches, 28, 31
Project Impact and, 71, 72
recommended mitigation actions,
    210
state government level
    recommendations, 217
Tulsa flood mitigation and
    prevention program, 90
Hydrology, and wildfires, 35

I

Ice caps, 3, 40
Ice storm of 2007, Tulsa, 104–105, 108
ICLEI - Local Governments for
    Sustainability, 208–209,
    225–226
Idaho, 38, 39
IFMI; *See* Red River International
    Flood Mitigation Initiative
Impact reduction framework,
    planning process, 21–22
Implementation, mitigation
    Project Impact goals, 75
    Project Impact pilot effort, 76, 78
    Red River IFMI, 157, 170
Incentive programs
    Berkeley case study, 117, 120
    FEMA Project Impact goals, 75
    National Flood Insurance Program,
        53, 54
    partnerships and program
        development, 212
    state government level
        recommendations, 217
Information, 209
Infrastructure, 209
    Berkeley case study, 116–117

mitigation and prevention program
    integration with, 212
In-kind contributions
    Seattle Project Impact, 174–175
    successful program requirements,
        205
Institute for Business and Home Safety,
    94, 101
Institutional capacity, 209
Insurance, 209
    and Coastal Barrier Resource System
        (CBRS) development, 58
    and coastal development, 28
    and development in floodplains, 43
    federal programs,
        recommendations for, 214
    NFIP; *See* National Flood Insurance
        Program
    Red River IFMI, 160, 171
    Tulsa Disaster-Resistant Business
        Council, 101
Insurance rates
    Napa River program and, 141
    Tulsa, 107
Integrated systems, 209
Interest groups, Project Impact
    development principles, 71
Intergovernmental Panel on Climate
    Change (IPCC), 1
    on adaptation, 8, 9
    Fourth Assessment Report: Climate
        Change 2007, 2, 3–5
    on mitigation, 7
    reports, 224–225
    worst case scenario developed by, 33
International Flood Mitigation Initiative
    (IFMI), 125–126, 149–170; *See
    also* Red River International
    Flood Mitigation Initiative
International Joint Commission (IJC),
    166
International Water Institute programs,
    169
Internet
    RiverWatch Project elements, 164